W9-ADX-830

The Oxford Book of Story Poems

The Oxford Book of Story Poems

Michael Harrison
and Christopher Stuart-Clark

Oxford University Press
Oxford New York Toronto

Contents

Fairy Story

I went into the wood one day
And there I walked and lost my way

When it was so dark I could not see
A little creature came to me

He said if I would sing a song
The time would not be very long

But first I must let him hold my hand tight
Or else the wood would give me a fright

I sang a song, he let me go
But now I am home again there is nobody I know.

Stevie Smith

Meet-on-the-Road

'Now pray, where are you going, child?'
  said Meet-on-the-Road.
'To school, sir, to school, sir,'
  said Child-as-it-Stood.
'What have you got in your basket, child?'
  said Meet-on-the-Road.
'My dinner, sir, my dinner, sir,'
  said Child-as-it-Stood.
'What have you for your dinner, child?'
  said Meet-on-the-Road.

'Some pudding, sir, some pudding, sir,'
  said Child-as-it-Stood.
'Oh then, I pray, give me a share,'
  said Meet-on-the-Road.
'I've little enough for myself, sir,'
  said Child-as-it-Stood.
'What have you got that cloak on for?'
  said Meet-on-the-Road.
'To keep the wind and the cold from me,'
  said Child-as-it-Stood.

'I wish the wind would blow through you,'
  Said Meet-on-the-Road.
'Oh, what a wish! Oh, what a wish!'
  said Child-as-it-Stood.
'Pray, what are those bells ringing for?'
  said Meet-on-the-Road.
'To ring bad spirits home again,'
  said Child-as-it-Stood.
'Oh, then I must be going, child!'
  said Meet-on-the-Road.
'So fare you well, so fare you well,'
  said Child-as-it-Stood.

Anon.

The Listeners

'Is there anybody there?' said the Traveller,
    Knocking on the moonlit door;
And his horse in the silence champed the grasses
    Of the forest's ferny floor;
And a bird flew up out of the turret,
    Above the Traveller's head:
And he smote upon the door again a second time;
    'Is there anybody there?' he said.
But no one descended to the Traveller;
    No head from the leaf-fringed sill
Leaned over and looked into his grey eyes,
    Where he stood perplexed and still.
But only a host of phantom listeners
    That dwelt in the lone house then
Stood listening in the quiet of the moonlight
    To that voice from the world of men:

Stood thronging the faint moonbeams on the dark stair,
That goes down to the empty hall,
Hearkening in an air stirred and shaken
By the lonely Traveller's call.
And he felt in his heart their strangeness,
Their stillness answering his cry,
While his horse moved, cropping the dark turf,
'Neath the starred and leafy sky;
For he suddenly smote on the door, even
Louder, and lifted his head:—
'Tell them I came, and no one answered,
That I kept my word,' he said.
Never the least stir made the listeners,
Though every word he spake
Fell echoing through the shadowiness of the still house
From the one man left awake:
Ay, they heard his foot upon the stirrup,
And the sound of iron on stone,
And how the silence surged softly backward,
When the plunging hoofs were gone.

Walter de la Mare

Jabberwocky

'Twas brillig, and the slithy toves
Did gyre and gimble in the wabe;
All mimsy were the borogroves,
And the mome raths outgrabe.

Beware the Jabberwock, my son!
The jaws that bite, the claws that catch!
Beware the Jubjub bird and shun
The frumious Bandersnatch!

He took his vorpal sword in hand:
Long time the manxome foe he sought—
So rested he by the Tumtum tree,
And stood awhile in thought.

And as in uffish thought he stood,
The Jabberwock, with eyes of flame,
Came whiffling through the tulgey wood,
And burbled as it came!

One two! One two! and through and through
The vorpal blade went snicker-snack!
He left it dead, and with its head
He went galumphing back.

'And hast thou slain the Jabberwock!
Come to my arms, my beamish boy!
O frabjous day! Callooh! Callay!'
He chortled in his joy.

'Twas brillig, and the slithy toves
Did gyre and gimble in the wabe;
All mimsy were the borogroves,
And the mome raths outgrabe.

Lewis Carroll

The Malfeasance

It was a dark, dank, dreadful night
And while millions were abed
The Malfeasance bestirred itself
And raised its ugly head.

The leaves dropped quietly in the night
In the sky Orion shone;
The Malfeasance bestirred itself
Then crawled around till dawn.

Taller than a chimney stack,
More massive than a church,
It slithered to the city
With a purpose and a lurch.

Squelch, squelch, the scaly feet
Flapped along the roads;
Nothing like it had been seen
Since a recent fall of toads.

Bullets bounced off the beast,
Aircraft made it grin,
Its open mouth made an eerie sound
Uglier than sin.

Still it floundered forwards,
Still the city reeled;
There was panic on the pavements,
Even policemen squealed.

Then suddenly someone suggested
(As the beast had done no harm)
It would be kinder to show it kindness,
Better to stop the alarm.

When they offered it refreshment
The creature stopped in its track;
When they waved a greeting to it
Steam rose from its back.

As the friendliness grew firmer
The problem was quietly solved:
Terror turned to triumph and
The Malfeasance dissolved.

And where it stood there hung a mist,
And in its wake a shining trail,
And the people found each other
And thereby hangs a tail.

Alan Bold

What has Happened to Lulu?

What has happened to Lulu, mother?
   What has happened to Lu?
There's nothing in her bed but an old rag-doll
   And by its side a shoe.

Why is her window wide, mother,
   The curtain flapping free,
And only a circle on the dusty shelf
   Where her money-box used to be?

Why do you turn your head, mother,
   And why do the tear-drops fall?
And why do you crumple that note on the fire
   And say it is nothing at all?

I woke to voices late last night,
   I heard an engine roar,
Why do you tell me the things I heard
   Were a dream and nothing more?

I heard somebody cry, mother,
   In anger or in pain,
But now I ask you why, mother,
   You say it was a gust of rain.

Why do you wander about as though
   You don't know what to do?
What has happened to Lulu, mother?
   What has happened to Lu?

Charles Causley

La Belle Dame Sans Merci

'O what can ail thee, knight-at-arms,
Alone and palely loitering?
The sedge is wither'd from the lake,
And no birds sing.

'O what can ail thee, knight-at-arms,
So haggard and so woe-begone?
The squirrel's granary is full,
And the harvest's done.

'I see a lily on thy brow,
With anguish moist and fever dew,
And on thy cheeks a fading rose
Fast withereth too.'

'I met a lady in the meads
Full beautiful—a faery's child;
Her hair was long, her foot was light,
And her eyes were wild.

'I made a garland for her head,
And bracelets too, and fragrant zone;
She look'd at me as she did love,
And made sweet moan.

'I set her on my pacing steed
And nothing else saw all day long;
For sideways would she bend, and sing
A faery's song.

'She found me roots of relish sweet,
And honey wild, and manna dew,
And sure in language strange she said,
"I love thee true."

'She took me to her elfin grot,
And there she wept, and sigh'd full sore,
And there I shut her wild, wild eyes
With kisses four.

‘And there she lulled me asleep,
And there I dream’d,—Ah! woe betide,
The latest dream I ever dream’d
On the cold hill side.

‘I saw pale kings, and princes too,
Pale warriors, death-pale were they all;
They cried—“La belle Dame sans Merci
Hath thee in thrall!”

‘I saw their starv’d lips in the gloam
With horrid warning gaped wide,
And I awoke, and found me here
On the cold hill side.

‘And this is why I sojourn here
Alone and palely loitering,
Though the sedge is wither’d from the lake
And no birds sing.’

John Keats

Annabel Lee

It was many and many a year ago,
    In a kingdom by the sea,
That a maiden there lived whom you may know
    By the name of Annabel Lee;
And this maiden she lived with no other thought
    Than to love and be loved by me.

I was a child and she was a child,
    In this kingdom by the sea;
But we loved with a love that was more than love—
    I and my Annabel Lee;
With a love that the wingèd seraphs of heaven
    Coveted her and me.

And this was the reason that, long ago,
    In this kingdom by the sea,
A wind blew out of a cloud, chilling
    My beautiful Annabel Lee;
So that her high born kinsmen came
    And bore her away from me,
To shut her up in a sepulchre
    In this kingdom by the sea.

The angels, not half so happy in heaven,
    Went envying her and me—
Yes!—that was the reason (as all men know,
    In this kingdom by the sea)
That the wind came out of the cloud by night,
    Chilling and killing my Annabel Lee.

But our love it was stronger by far than the love
    Of those who were older than we—
    Of many far wiser than we—
And neither the angels in heaven above,
    Nor the demons down under the sea,
Can ever dissever my soul from the soul
    Of the beautiful Annabel Lee.

For the moon never beams without bringing me dreams
    Of the beautiful Annabel Lee;
And the stars never rise but I feel the bright eyes
    Of the beautiful Annabel Lee;
And so, all the night-tide, I lie down by the side
Of my darling—my darling—my life and my bride,
    In the sepulchre there by the sea,
    In her tomb by the sounding sea.

Edgar Allan Poe

As Lucy Went A-Walking

As Lucy went a-walking one morning cold and fine,
There sate three crows upon a bough, and three times three are nine:
Then 'O!' said Lucy, in the snow, 'it's very plain to see
A witch has been a-walking in the fields in front of me.'

Then stept she light and heedfully across the frozen snow,
And plucked a bunch of elder-twigs that near a pool did grow:
And, by and by, she comes to seven shadows in one place
Stretched black by seven poplar-trees against the sun's bright face.

She looks to left, she looks to right, and in the midst she sees
A little pool of water clear and frozen 'neath the trees;
Then down beside its margent in the crusty snow she kneels,
And hears a magic belfry, ringing with sweet bells.

Clear rang the faint far merry peal, then silence on the air,
And icy-still the frozen pool and poplars standing there:
Then soft, as Lucy turned her head and looked along the snow
She sees a witch—a witch she sees, come frisking to and fro.

Her scarlet, buckled shoes they clicked, her heels a-twinkling high;
With mistletoe her steeple-hat bobbed as she capered by;
But never a dint, or mark, or print, in the whiteness for to see,
Though danced she light, though danced she fast, though danced she lissomely.

It seemed 'twas diamonds in the air, or tiny flakes of frost;
It seemed 'twas golden smoke around, or sunbeams lightly tossed;
It seemed an elfin music like to reeds' and warblers' rose:
'Nay!' Lucy said, 'it is the wind that through the branches flows.'

And as she peeps, and as she peeps, 'tis no more one, but three,
And eye of bat, and downy wing of owl within the tree,
And the bells of that sweet belfry a-pealing as before,
And now it is not three she sees, and now it is not four.

'O! who are ye,' sweet Lucy cries, 'that in a dreadful ring,
All muffled up in brindled shawls, do caper, frisk, and spring?'
'A witch and witches, one and nine,' they straight to her reply,
And looked upon her narrowly, with green and needle eye.

Then Lucy sees in clouds of gold sweet cherry-trees upgrow,
And bushes of red roses that bloomed above the snow;
She smells all faint the almond-boughs blowing so wild and fair,
And doves with milky eyes ascend fluttering in the air.

Clear flowers she sees, like tulip buds, go floating by like birds,
With wavering tips that warbled sweetly strange enchanted words;
And as with ropes of amethyst the twigs with lamps were hung,
And clusters of green emeralds like fruit upon them clung.

'O witches nine, ye dreadful nine, O witches three times three!
Whence come these wondrous things that I this Christmas morning see?'
But straight, as in a clap, when she of 'Christmas' says the word,
Here is the snow, and there the sun, but never bloom nor bird;

Nor warbling flame, nor gloaming-rope of amethyst there shows,
Nor bunches of green emeralds, nor belfry, well, and rose,
Nor cloud of gold, nor cherry-tree, nor witch in brindled shawl,
But like a dream which vanishes, so vanished were they all.

When Lucy sees, and only sees three crows upon a bough,
And earthly twigs, and bushes hidden white in driven snow,
Then 'O!' said Lucy, 'three times three are nine—I plainly see
Some witch has been a-walking in the fields in front of me.'

Walter de la Mare

The Hairy Toe

Once there was a woman went out to pick beans,
and she found a Hairy Toe.
She took the Hairy Toe home with her,
and that night, when she went to bed,
the wind began to moan and groan.
Away off in the distance
she seemed to hear a voice crying,
'Where's my Hair-r-ry To-o-e?
Who's got my Hair-r-ry To-o-oe?'

The woman scrooched down,
way down under the covers,
and about that time
the wind appeared to hit the house,

smoosh,

and the old house creaked and cracked
like something was trying to get in.
The voice had come nearer,
almost at the door now,
and it said,
'Where's my Hair-r-ry To-o-oe?
Who's got my Hair-r-ry To-o-oe?'

The woman scrooched further down
under the covers
and pulled them tight around her head.

The wind growled around the house
like some big animal
and r-r-um-umbled
over the chimbley.
All at once she heard the door cr-r-a-ack
and Something slipped in
and began to creep over the floor.

The floor went
cre-e-eak, cre-e-eak
at every step that thing took towards her bed.
The woman could almost feel
it bending over her bed.
There in an awful voice it said:
'Where's my Hair-r-ry To-o-oe?
Who's got my Hair-r-ry To-o-oe?
You've got it!'

Traditional

Life Story

Once — but no matter when —
   There lived — no matter where —
A man, whose name — but then
   I need not that declare.

He — well, he had been born,
   And so he was alive;
His age — I details scorn —
   Was somethingty and five.

He lived — how many years
   I truly can't decide;
But this one fact appears
   He lived — until he died.

'He died,' I have averred,
   But cannot prove 'twas so,
But that he was interred,
   At any rate, I know.

I fancy he'd a son,
   I hear he had a wife:
Perhaps he'd more than one,
   I know not, on my life!

But whether he was rich,
   Or whether he was poor,
Or neither — both — or which,
   I cannot say, I'm sure.

I can't recall his name,
   Or what he used to do:
But then — well, such is fame!
   'Twill so serve me and you.

And that is why I thus,
   About this unknown man
Would fain create a fuss,
   To rescue, if I can

From dark oblivion's blow,
   Some record of his lot:
But, ah! I do not know
   Who — where — when — why — or what.

Anon.

Storytime

Once upon a time, children,
there lived a fearsome dragon.

Please, miss,
Jamie's made a dragon.
Out in the sandpit.

Lovely, Andrew.
Now this dragon
had enormous red eyes
and a swirling, whirling tail . . .

Jamie's dragon's got
yellow eyes, miss.

Lovely, Andrew.
Now this dragon was
as wide as a horse
as green as the grass
as tall as a house . . .

Jamie's would JUST fit
in our classroom, miss!

But he was a very friendly dragon . . .

Jamie's dragon ISN'T, miss.
He eats people, miss.
Especially TEACHERS,
Jamie said.

Very nice, Andrew!
Now one day, children,
this enormous dragon
rolled his red eye,
whirled his swirly green tail
and set off to find . . .

His dinner, miss!
Because he was hungry, miss!

Thank you, Andrew.
He rolled his red eye,
whirled his green tail,
and opened his wide, wide mouth
until

Please, miss,
I did try to tell you, miss!

Judith Nicholls

Jim Who Ran Away from his Nurse, and was Eaten by a Lion

There was a boy whose name was Jim;
His friends were very good to him.
They gave him tea, and cakes, and jam,
And slices of delicious ham,
And chocolate with pink inside,
And little tricycles to ride,
And read him stories through and through,
And even took him to the Zoo—
But there it was the dreadful fate
Befell him, which I now relate.

You know—at least you *ought* to know,
For I have often told you so—
That children never are allowed
To leave their nurses in a crowd;
Now this was Jim's especial foible,
He ran away when he was able,
And on this inauspicious day
He slipped his hand and ran away!
He hadn't gone a yard when—Bang!
With open jaws, a lion sprang,
And hungrily began to eat
The boy: beginning at his feet.

Now, just imagine how it feels
When first your toes and then your heels,
And then by gradual degrees,
Your shins and ankles, calves and knees,
Are slowly eaten, bit by bit.
No wonder Jim detested it!
No wonder that he shouted 'Hi!'
The honest keeper heard his cry,
Though very fat he almost ran
To help the little gentleman.
'Ponto!' he ordered as he came
(For Ponto was the lion's name),
'Ponto!' he cried, with angry frown
'Let go, Sir! Down, Sir! Put it down!'

The lion made a sudden stop,
He let the dainty morsel drop,
And slunk reluctant to his cage,
Snarling with disappointed rage.
But when he bent him over Jim,
The honest keeper's eyes were dim.
The lion having reached his head,
The miserable boy was dead!

When Nurse informed his parents, they
Were more concerned than I can say:—
His Mother, as she dried her eyes,
Said, 'Well—it gives me no surprise,
He would not do as he was told!'
His Father, who was self-controlled,
Bade all the children round attend
To James's miserable end,
And always keep a-hold of Nurse
For fear of finding something worse.

Hilaire Belloc

What Happened to Miss Frugle

Stern Miss Frugle always said
To Peter and his sister
'After school you'll stay behind
If you so much as whisper.'

Then one winter afternoon
While skating on thin ice
The children saw it crack and Miss
Frugle vanish in a trice.

People wondered where she'd gone,
But no one really missed her,
And she was never found because
Peter and his sister

didn't so much as whisper
didn't so much as whisper.

Brian Patten

Ballad

A knight went down to the river's rim
And saw a nymph glance back at him.

'The river's daughter herself I am,'
And into his waiting arms she swam.

Then summer's ardour stretched out and loved
As the cool water beneath him moved.

But when their loving force was spent
The nymph dissolved in her element.

And after many great vows and tall
The knight rode away by the steep cliff wall.

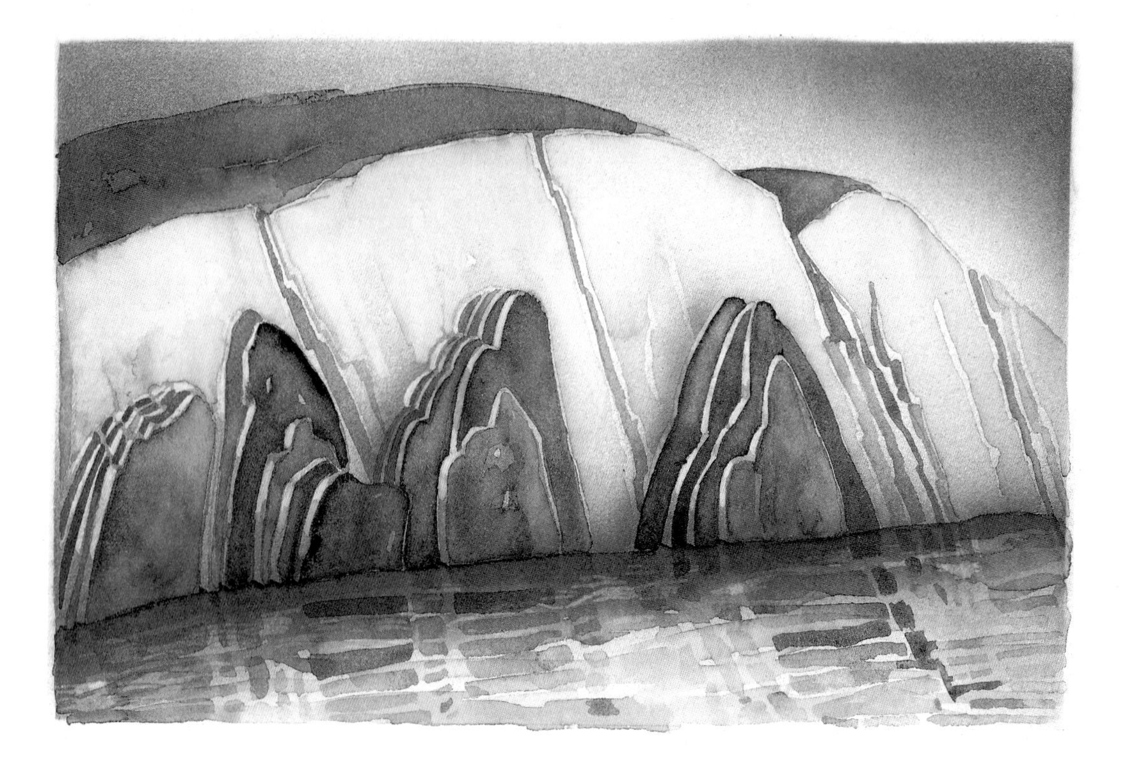

'I'll come this way again,' he said,
'And marry the nymph of the river bed.'

The knight rode away and remembered his cares:
'First I must settle my weighty affairs,

Instruct the steward of my estate,
And fix the bolt on the garden gate;

Must pay my men and harvest the grain
Before I come back to the river again.'

The gate is fixed, the grain is sold,
The weather grows bleak, the year turns cold,

And part of the river is frozen over
As the nymph awaits the return of her lover.

The knight is having a Christmas fling
And tells his heart it has time until spring.

He dances with his neighbour's daughter,
Who's as gold to silver, as sun to water.

So he forgets the song of the river
And swears new love for ever and ever.

Now handsome knight and lovely bride
In the month of May to the river ride.

Together with many guests of rank
They ride along to the river-bank.

And down the hill clip-clops their train
Who will not return that way again.

'My bride, my bride, why do we go
To where the sullen waters flow?'

But the bride shines blonde as the midday sun
And she and the knight and the guests are undone.

They ride past the cliff where the waters moan
And are turned forever to rock and stone.

Cold as rock and as still and stiff
They now form part of the river's cliff:

The guests on horseback, the monk, the bride
And the faithless bridegroom at her side.

But the nymph looks up and repents her deed:
'I should have allowed them to pass,' she said,

'For, turned to stone, forever they'll be
A sorrow and a reproach to me.'

The centuries pass, the pleasure-boats go
Carrying sightseers to and fro

Who in their time will turn as stiff:
For all, all is water beneath the cliff.

Gerda Mayer

After Ever Happily

or The Princess and the Woodcutter*

And they both lived happily ever after . . .
The wedding was held in the palace. Laughter
Rang to the roof as a loosened rafter
Crashed down and squashed the chamberlain flat —
And how the wedding guests chuckled at that!
'You, with your horny indelicate hands,
Who drop your haitches and call them 'ands,
Who cannot afford to buy her a dress,
How dare you presume to pinch our princess —
Miserable woodcutter, uncombed, unwashed!'
Were the chamberlain's words (before he was squashed).
'Take her,' said the Queen, who had a soft spot
For woodcutters. 'He's strong and he's handsome. Why not?'
'What rot!' said the King, but he dare not object;
The Queen wore the trousers—that's as you'd expect.
Said the chamberlain, usually meek and inscrutable,
'A princess and a woodcutter? The match is unsuitable.'
Her dog barked its welcome again and again,
As they splashed to the palace through puddles of rain.
And the princess sighed, 'Till the end of my life!'
'Darling,' said the woodcutter, 'will you be my wife?'
He knew all his days he could love no other,
So he nursed her to health with some help from his mother,
And lifted her, horribly hurt, from her tumble.
A woodcutter, watching, saw the horse stumble.
As she rode through the woods, a princess in her prime
On a dapple-grey horse . . . Now, to finish my rhyme,
I'll start it properly: Once upon a time —

* This is a love story from the Middle Ages.
The poet obviously knew his subject-matter
backwards.

Ian Serraillier

Three Wise Old Women

Three wise old women were they, were they,
Who went to walk on a winter day:
One carried a basket to hold some berries,
One carried a ladder to climb for cherries,
The third, and she was the wisest one,
Carried a fan to keep off the sun.

But they went so far, and they went so fast,
They quite forgot their way at last,
So one of the wise women cried in a fright,
'Suppose we should meet a bear tonight!
Suppose he should eat me!' 'And me!!' 'And me!!!'
'What is to be done?' cried all the three.

'Dear, dear!' said one, 'we'll climb a tree,
There out of the way of the bears we'll be.'
But there wasn't a tree for miles around;
They were too frightened to stay on the ground,
So they climbed their ladder up to the top,
And sat there screaming, 'We'll drop! We'll drop!'

But the wind was strong as wind could be,
And blew their ladder right out to sea;
So the three wise women were all afloat
In a leaky ladder instead of a boat,
And every time the waves rolled in,
Of course the poor things were wet to the skin.

Then they took their basket, the water to bale,
They put up their fan instead of a sail:
But what became of the wise women then
Whether they ever sailed home again,
Whether they saw any bears, or no,
You must find out, for I don't know.

Elizabeth T. Corbett

The Old Woman and the Sandwiches

I met a wizened wood-woman
    Who begged a crumb of me.
Four sandwiches of ham I had:
    I gave her three.

'Bless you, thank you, kindly Miss,
    Shall be rewarded well—
Three everlasting gifts, whose value
    None can tell.'

'Three wishes?' out I cried in glee—
    'No, gifts you may not choose:
A flea and gnat to bite your back
    And gravel in your shoes.'

Libby Houston

The Nose

(after Gogol)

The nose went away by itself
in the early morning
while its owner was asleep.
It walked along the road
sniffing at everything.

It thought: I have a personality of my own.
Why should I be attached to a body?
I haven't been allowed to flower.
So much of me has been wasted.

And it felt wholly free.
It almost began to dance
The world was so full of scents
it had had no time to notice,

when it was attached to a face
weeping, being blown,
catching all sorts of germs
and changing colour.

But now it was quite at ease
bowling merrily along
like a hoop or a wheel,
a factory packed with scent.

And all would have been well
but that, round about evening,
having no eyes for guides,
it staggered into the path
of a mouth, and it was gobbled
rapidly like a sausage
and chewed by great sour teeth—
and that was how it died.

Iain Crichton Smith

The Walrus and the Carpenter

The sun was shining on the sea,
Shining with all his might:
He did his very best to make
The billows smooth and bright—
And this was odd, because it was
The middle of the night.

The moon was shining sulkily,
Because she thought the sun
Had got no business to be there
After the day was done—
'It's very rude of him,' she said,
'To come and spoil the fun!'

The sea was wet as wet could be,
The sands were dry as dry.
You could not see a cloud, because
No cloud was in the sky:
No birds were flying overhead—
There were no birds to fly.

The Walrus and the Carpenter
Were walking close at hand;
They wept like anything to see
Such quantities of sand;
'If this were only cleared away,'
They said, 'it *would* be grand!'

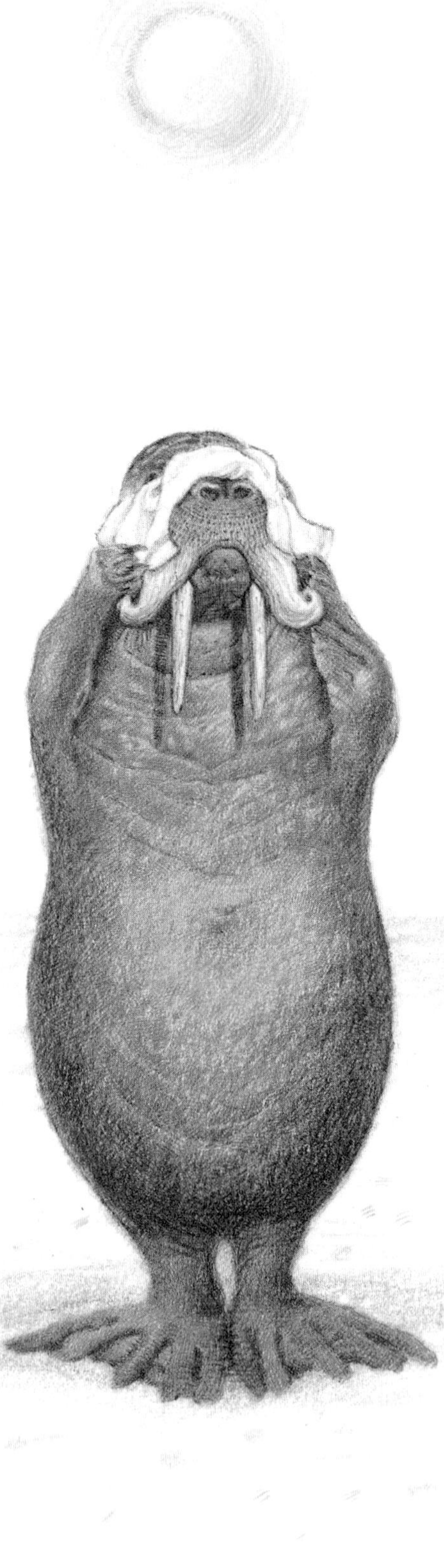

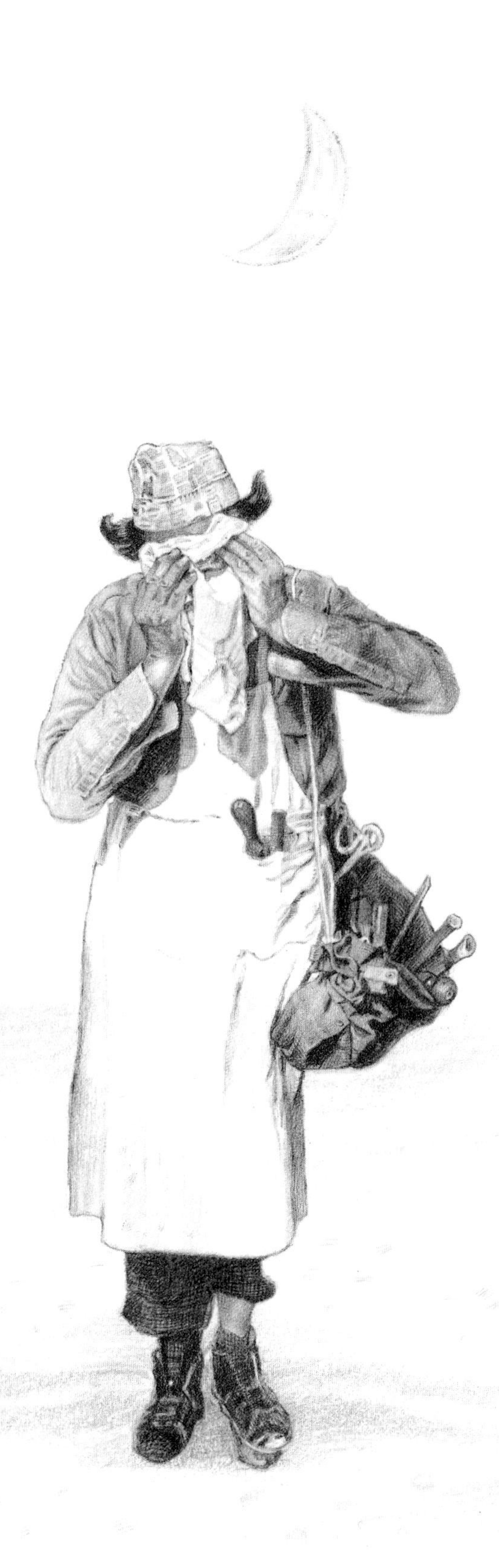

'If seven maids with seven mops
Swept it for half a year,
Do you suppose,' the Walrus said,
'That they could get it clear?'
'I doubt it,' said the Carpenter,
And shed a bitter tear.

'O Oysters, come and walk with us!'
The Walrus did beseech.
'A pleasant walk, a pleasant talk,
Along the briny beach:
We cannot do with more than four,
To give a hand to each.'

The eldest Oyster looked at him,
But never a word he said:
The eldest Oyster winked his eye,
And shook his heavy head—
Meaning to say he did not choose
To leave the oyster-bed.

But four young Oysters hurried up,
All eager for the treat:
Their coats were brushed, their faces washed,
Their shoes were clean and neat—
And this was odd, because, you know,
They hadn't any feet.

Four other Oysters followed them,
And yet another four;
And thick and fast they came at last,
And more, and more, and more—
All hopping through the frothy waves,
And scrambling to the shore.

The Walrus and the Carpenter
Walked on a mile or so,
And then they rested on a rock
Conveniently low:
And all the little Oysters stood
And waited in a row.

'The time has come,' the Walrus said,
'To talk of many things:
Of shoes—and ships—and sealing wax—
Of cabbages—and kings—
And why the sea is boiling hot—
And whether pigs have wings.'

'But wait a bit,' the Oysters cried,
'Before we have our chat;
For some of us are out of breath,
And all of us are fat!'
'No hurry!' said the Carpenter.
They thanked him much for that.

'A loaf of bread,' the Walrus said,
  'Is what we chiefly need;
Pepper and vinegar besides
  Are very good indeed—
Now, if you're ready, Oysters dear,
  We can begin to feed.'

'But not on us,' the Oysters cried,
  Turning a little blue.
'After such kindness that would be
  A dismal thing to do!'
'The night is fine,' the Walrus said,
  'Do you admire the view?'

'It was so kind of you to come,
  And you are very nice!'
The Carpenter said nothing but,
  'Cut us another slice.
I wish you were not quite so deaf—
  I've had to ask you twice!'

'It seems a shame,' the Walrus said,
  'To play them such a trick.
After we've brought them out so far
  And made them trot so quick!'
The Carpenter said nothing but,
  'The butter's spread too thick!'

‘I weep for you,’ the Walrus said,
    ‘I deeply sympathize.’
With sobs and tears he sorted out
    Those of the largest size,
Holding his pocket-handkerchief
    Before his streaming eyes.

‘O Oysters,’ said the Carpenter,
    ‘You’ve had a pleasant run!
Shall we be trotting home again?’
    But answer came there none—
And this was scarcely odd, because
    They’d eaten every one.

Lewis Carroll

One Winter Night in August

One winter night in August
While the larks sang in their eggs,
A barefoot boy with shoes on
Stood kneeling on his legs.

At ninety miles an hour
He slowly strolled to town
And parked atop a tower
That had just fallen down.

He asked a kind old policeman
Who bit small boys in half,
‘Officer, have you seen my pet
Invisible giraffe?’

‘Why, sure, I haven’t seen him.’
The cop smiled with a sneer.
‘He was just here tomorrow
And he rushed right back next year.

'Now, boy, come be arrested
For stealing frozen steam!'
And whipping out his pistol,
He carved some hot ice cream.

Just then a pack of dogfish
Who roam the desert snows
Arrived by unicycle
And shook the policeman's toes.

They cried, 'Congratulations,
Old dear! Surprise, surprise!
You raced the worst, so you came in first
And you didn't win any prize!'

Then turning to the boyfoot bear,
They yelled, 'He's overheard
What we didn't say to the officer!
(We never said one word!)

'Too bad, boy, we must turn you
Into a loathsome toad!
Now shut your ears and listen,
We're going to explode!'

But then, with an awful holler
That didn't make a peep,
Our ancient boy (aged seven)
Woke up and went to sleep.

X. J. Kennedy

Here is the News

In Manchester today a man was seen
with hair on top of his head.
Over now straightway to our Northern correspondent:
Hugh Snews.

'It's been a really incredible day
here in Manchester. Scenes like this
have been seen here everyday
for years and years. It's now quite certain
no one will be saying anything about this
for months to come. One eyewitness said so.
"Are you sure?" I said.
She said: "No."
Back to you in London.'

All round the world,
newspapers, radio and television
have taken no notice of this story
and already a Prime Minister
has said nothing about it at all.
What next?
Rumour McRumourbungle,
Expert expert in expert experts?

'I doubt it. I doubt whether
anyone *will* doubt it—but I do.'

'What?'

'Doubt it.'

Thank you, Rumour McRumourbungle.
But how did it all begin?
As dawn broke in Manchester it soon became clear.
It's quite likely there was a lot of air in the air.
An hour after a few minutes had gone,
a couple of seconds passed
and a minute later at 12.15
it was a quarter past twelve.

Suddenly from across the other side of the road,
on the side facing this side,
there was the same road from the other side.
This side was now facing that side
and the road on that other side
was still opposite this.
Then – it happened.
There is no question of this.
In fact – no one has questioned it at all.
Further proof of this comes from the police
who say that a woman held for questioning
was released immediately
because she didn't know any of the answers
to the questions that no one asked her . . .

So –
it's something of a mystery.
Yes –
it's a mysterious thing to some
and there are some who think
it could
in a mysterious way
be nothing at all.

Michael Rosen

Welsh Incident

'But that was nothing to what things came out
From the sea-caves of Criccieth yonder.'
'What were they? Mermaids? dragons? ghosts?'
'Nothing at all of any things like that.'
'What were they, then?'
'All sorts of queer things,
Things never seen or heard or written about,
Very strange, un-Welsh, utterly peculiar
Things. Oh, solid enough they seemed to touch,
Had anyone dared it. Marvellous creation,
All various shapes and sizes and no sizes,
All new, each perfectly unlike his neighbour,
Though all came moving slowly out together.'
'Describe just one of them.'
'I am unable.'
'What were their colours?'
'Mostly nameless colours,
Colours you'd like to see; but one was puce
Or perhaps more like crimson, but not purplish.
Some had no colour.'
'Tell me, had they legs?'
'Not a leg or foot among them that I saw.'
'But did these things come out in any order?
What o'clock was it? What was the day of the week?
Who else was present? How was the weather?'
'I was coming to that. It was half-past three
On Easter Tuesday last. The sun was shining.
The Harlech Silver Band played *Marchog Jesu*
On thirty-seven shimmering instruments,
Collecting for Carnarvon's (Fever) Hospital Fund.
The populations of Pwllheli, Criccieth,
Portmadoc, Borth, Tremadoc, Penrhyndeudraeth,
Were all assembled. Criccieth's mayor addressed them
First in good Welsh and then in fluent English,
Twisting his fingers in his chain of office,
Welcoming the things. They came out on the sand,
Not keeping time to the band, moving seaward
Silently at a snail's pace. But at last
The most odd, indescribable thing of all

Which hardly one man there could see for wonder
Did something recognizably a something.'
'Well, what?'
'It made a noise.'
'A frightening noise?'
'No, no.'
'A musical noise? A noise of scuffling?'
'No, but a very loud, respectable noise—
Like groaning to oneself on Sunday morning
In Chapel, close before the second psalm.'
'What did the mayor do?'
'I was coming to that.'

Robert Graves

Journey

I am the acorn
that grew the oak
that gave the plank
the Vikings took
to make a boat
to sail them out
across the seas
to England.

Judith Nicholls

A Speck Speaks

About ten million years ago
I was a speck of rock in a vast black rock.
My address was:
    Vast Black Rock,
    Near Italy,
    Twelve Metres Under
The Mediterranean Sea.

The other specks and I
Formed an impressive edifice—
Bulbously curving at the base
With rounded caves
And fun tunnels for the fish,
Romantically jagged at the top.

Life, for us specks, was uneventful—
One for all, welded together
In the cool, salty wet.
What more could specks
Expect?

Each year a few of us were lost,
Scrubbed from the edges of the rock
By the washerwoman waters
Which smoothed our base, whittled our cornices
And sharpened our pinnacles.
As the rock slowly shed skin-thin layers
It was my turn to be exposed
Among the packed grit of its surface,
(Near the tip of the fifty-ninth spire
From the end of the eastern outcrop).

One day, it was a Wednesday I remember,
A scampi flicked me off my perch
Near the vast black rock's peak
And I was scurried down
Long corridors of currents
Until a wave caught me in its mouth
And spat me out on—
What?

A drying stretch
Of yellow, white, black, red and transparent specks,
Billions of particles,
Loosely organized in bumps and dips;
Quite unlike the tight hard group
Which I belonged to in the good old rock.
Heat banged down on us all day long.
Us? I turned to the speck next to me,
A lumpish red fellow who'd been washed off a brick.

'I'm new here,' I confessed,
'What are we supposed to be?'
He bellowed back—
(But the bellow of a speck
Is less than the whispering of ants)—
'We're grains now, grains of sand,
And this society is called Beach.'

'Beach?' I said. 'What are we grains supposed to do?'
'Just stray around, lie loose,
Go with the wind, go with the sea
And sink down when you're trodden on.'

'Don't know if I can manage that.
Used to belong to Vast Black Rock
And we all stuck together.'

'Give Beach a try,' said the red grain.
Well, there was no alternative.

Many eras later
I was just beginning to feel
Part of Beach, that slow-drifting,
Slow-shifting, casual community,
When I was shovelled up
With a ton of fellow grains,
Hoisted into a lorry, shaken down a road,
Washed, sifted and poured in a machine
Hotter than the sunshine.

When they poured me out, life had changed again.
My mates and I swam in a molten river
Down into a mould.
White-hot we were, then red, then
Suddenly cold
And we found ourselves merged
Into a tall, circular tower,
Wide at the bottom, narrow at the top
What's more, we'd all turned green as sea-weed.
Transparent green.
We had become—a wine bottle.

In a few flashes of time
We'd been filled with wine,
Stoppered, labelled, bumped to a shop,
Stood in a window, sold, refrigerated,
Drained by English tourists,
Transmogrified into a lampstand,
Smashed by a four-year-old called Tarquin,
Swept up, chucked in the garbage, hauled away,
Dumped and bulldozed into the sea.

Now the underwater years sandpaper away
My shield-shaped fragment of bottle.
So one day I shall be a single grain again,
A single grain of green, transparent glass.

When that day comes
I will transmit a sub-aquatic call
To all green specks of glass
Proposing that we form
A Vast Green Rock of Glass
Near Italy
Twelve Metres Under
The Mediterranean Sea.

Should be pretty spectacular
In about ten million years.

All being well.

Adrian Mitchell

The Forsaken Merman

Come, dear children, let us away;
Down and away below!
Now my brothers call from the bay,
Now the great winds shoreward blow,
Now the salt tides seaward flow;
Now the wild white horses play,
Champ and chafe and toss in the spray.
Children dear, let us away!
This way, this way!

Call her once before you go—
Call once yet!
In a voice that she will know:
'Margaret! Margaret!'
Children's voices should be dear
(Call once more) to a mother's ear;
Children's voices, wild with pain—
Surely she will come again!
Call her once and come away;
This way, this way!
'Mother dear, we cannot stay!
The wild white horses foam and fret.'
Margaret! Margaret!

Come, dear children, come away down;
Call no more!
One last look at the white-wall'd town,
And the little grey church on the windy shore;
Then come down!
She will not come though you call all day;
Come away, come away!

Children dear, was it yesterday
We heard the sweet bells over the bay?
In the caverns where we lay,
Through the surf and through the swell,
The far-off sound of a silver bell?
Sand-strewn caverns, cool and deep,
Where the winds are all asleep;
Where the spent lights quiver and gleam,
Where the salt weed sways in the stream,

Where the sea-beasts, ranged all round,
Feed in the ooze of their pasture-ground;
Where the sea-snakes coil and twine,
Dry their mail and bask in the brine;
Where great whales come sailing by,
Sail and sail, with unshut eye,
Round the world for ever and aye?
When did music come this way?
Children dear, was it yesterday?

Children dear, was it yesterday
(Call yet once) that she went away?
Once she sate with you and me,
On a red gold throne in the heart of the sea,
And the youngest sate on her knee.
She comb'd its bright hair, and she tended it well,
When down swung the sound of a far-off bell.
She sigh'd, she look'd up through the clear green sea;
She said: 'I must go, for my kinsfolk pray
In the little grey church on the shore today.
'Twill be Easter-time in the world—ah me!
And I lose my poor soul, Merman! here with thee.'
I said: 'Go up, dear heart, through the waves;
Say thy prayer, and come back to the kind sea-caves!'
She smiled, she went up through the surf in the bay.
Children dear, was it yesterday?

Children dear, were we long alone?
'The sea grows stormy, the little ones moan;
Long prayers,' I said, 'in the world they say;
Come!' I said; and we rose through the surf in the bay.
We went up the beach, by the sandy down
Where the sea-stocks bloom, to the white-wall'd town;
Through the narrow paved streets, where all was still,
To the little grey church on the windy hill.
From the church came a murmur of folk at their prayers,
But we stood without in the cold blowing airs.
We climb'd on the graves, on the stones worn with rains,
And we gazed up the aisle through the small leaded panes.
She sate by the pillar; we saw her clear:
'Margaret, hist! come quick, we are here!
Dear heart,' I said, 'we are long alone;
The sea grows stormy, the little ones moan.'
But, ah, she gave me never a look,
For her eyes were seal'd to the holy book!
Loud prays the priest; shut stands the door.
Come away, children, call no more!
Come away, come down, call no more!

Down, down, down!
Down to the depths of the sea!
She sits at her wheel in the humming town,
Singing most joyfully.
Hark what she sings: 'O joy, O joy,
For the humming street, and the child with its toy!
For the priest, and the bell, and the holy well;
For the wheel where I spun,
And the blessed light of the sun!'
And so she sings her fill,
Singing most joyfully,
Till the spindle drops from her hand,
And the whizzing wheel stands still.
She steals to the window, and looks at the sand,
And over the sand at the sea;
And her eyes are set in a stare;
And anon there breaks a sigh,
And anon there drops a tear,
From a sorrow-clouded eye,
And a heart sorrow-laden,
A long, long sigh;
For the cold strange eyes of a little Mermaiden
And the gleam of her golden hair.

Come away, away children;
Come children, come down!
The hoarse wind blows coldly;
Lights shine in the town.
She will start from her slumber
When gusts shake the door;
She will hear the winds howling,
Will hear the waves roar.
We shall see, while above us
The waves roar and whirl,
A ceiling of amber,
A pavement of pearl.
Singing: 'Here came a mortal,
But faithless was she!
And alone dwell for ever
The kings of the sea.'

But, children, at midnight,
When soft the winds blow,
When clear falls the moonlight,
When spring-tides are low;
When sweet airs come seaward
From heaths starr'd with broom,
And high rocks throw mildly
On the blanch'd sands a gloom;
Up the still, glistening beaches,
Up the creeks we will hie,
Over banks of bright seaweed
The ebb-tide leaves dry.
We will gaze, from the sand-hills,
At the white, sleeping town;
At the church on the hill-side—
And then come back down.
Singing: 'There dwells a loved one,
But cruel is she!
She left lonely for ever
The kings of the sea.'

Matthew Arnold

The Inchcape Rock

1. *The Inchcape Bell*

No stir in the air, no stir in the sea,
The ship was still as she could be,
Her sails from heaven received no motion,
Her keel was steady in the ocean.

Without either sign or sound of their shock
The waves flowed over the Inchcape Rock;
So little they rose, so little they fell,
They did not move the Inchcape Bell.

The Abbot of Aberbrothok
Had placed that bell on the Inchcape Rock;
On a buoy in the storm it floated and swung,
And over the waves its warning rung.

When the Rock was hid by the surge's swell,
The mariners heard the warning bell;
And then they knew the perilous Rock
And blest the Abbot of Aberbrothok.

2. *Sir Ralph the Rover's Wicked Deed*

The sun in heaven was shining gay,
All things were joyful on that day;
The sea-birds screamed as they wheeled round,
And there was joyaunce in their sound.

The buoy of the Inchcape Bell was seen
A darker speck on the ocean green;
Sir Ralph the Rover walked his deck,
And he fixed his eye on the darker speck.

He felt the cheering power of spring;
It made him whistle, it made him sing;
His heart was mirthful to excess.
But the Rover's mirth was wickedness.

His eye was on the Inchcape float;
Quoth he, 'My men, put out the boat,
And row me to the Inchcape Rock,
And I'll plague the Abbot of Aberbrothok.'

The boat is lowered, the boatmen row,
And to the Inchcape Rock they go;
Sir Ralph bent over from the boat,
And he cut the Bell from the Inchcape float.

Down sunk the Bell with a gurgling sound,
The bubbles rose and burst around;
Quoth Sir Ralph, 'The next who comes to the Rock
Won't bless the Abbot of Aberbrothok.'

3. *Sir Ralph the Rover's Return*

Sir Ralph the Rover sailed away,
He scoured the seas for many a day;
And now grown rich with plundered store,
He steers his course for Scotland's shore.

So thick a haze o'erspreads the sky
They cannot see the sun on high;
The wind hath blown a gale all day,
At evening it hath died away.

On the deck the Rover takes his stand,
So dark it is they see no land.
Quoth Sir Ralph, 'It will be lighter soon
For there is the dawn of the rising Moon.'

'Canst hear,' said one, 'the breakers roar?
For methinks we should be near the shore.'
'Now where we are I cannot tell.
But I wish I could hear the Inchcape Bell.'

They hear no sound, the swell is strong:
Though the wind hath fallen, they drift along.
Till the vessel strikes with a shivering shock.—
'Oh, Christ! It is the Inchcape Rock!'

Sir Ralph the Rover tore his hair;
He cursed himself in his despair;
The waves rush in on every side,
The ship is sinking beneath the tide.

But even in his dying fear
One dreadful sound could the Rover hear,
A sound as if with the Inchcape Bell
The Devil below was ringing his knell.

Robert Southey

Mary Celeste

Only the wind sings
in the riggings,
the hull creaks a lullaby;
a sail lifts gently
like a message
pinned to a vacant sky.
The wheel turns
over bare decks,
shirts flap on a line;
only the song of the lapping waves
beats steady time . . .

First mate,
off-duty from
the long dawn watch, begins
a letter to his wife, daydreams
of home.

The Captain's wife is late;
the child did not sleep
and breakfast has passed . . .
She, too, is missing home;
sits down at last to eat,
but can't quite force
the porridge down.
She swallows hard,
slices the top from her egg.

The second mate
is happy.
A four-hour sleep,
full stomach
and a quiet sea
are all he craves.
He has all three.

Shirts washed and hung, beds
made below, decks done, the boy
stitches a torn sail.

The Captain
has a good ear for a tune;
played his child to sleep
on the ship's organ.
Now, music left,
he checks his compass,
lightly tips the wheel,
hopes for a westerly.
Clear sky, a friendly sea,
fair winds for Italy.

The child now sleeps, at last,
head firmly pressed into her pillow
in a deep sea-dream.

Then why are the gulls wheeling
like vultures in the sky?
Why was the child snatched
from her sleep? What drew
the Captain's cry?

Only the wind replies
in the rigging,
and the hull creaks and sighs;
a sail spells out its message
over silent skies.
The wheel still turns
over bare decks,
shirts blow on the line;
the siren-song of lapping waves
still echoes over time.

Judith Nicholls

Sir Patrick Spens

I *The Sailing*

The king sits in Dunfermline town
Drinking the blude-red wine;
'O where will I get a skeely skipper
To sail this new ship o' mine?'

O up and spake an eldern knight,
Sat at the king's right knee:
'Sir Patrick Spens is the best sailor
That ever sail'd the sea.'

Our king has written a broad letter,
And seal'd it with his hand,
And sent it to Sir Patrick Spens,
Was walking on the strand.

'To Noroway, to Noroway,
To Noroway o'er the foam;
The king's daughter o' Noroway,
'Tis thou must bring her home.'

The first word that Sir Patrick read
So loud, loud laugh'd he;
The next word that Sir Patrick read
The tear blinded his e'e.

'O who is this has done this deed
And told the king o' me,
To send us out, at this time o' year,
To sail upon the sea?

'Be it wind, be it wet, be it hail, be it sleet,
Our ship must sail the foam;
The king's daughter o' Noroway,
'Tis we must fetch her home.'

They hoisted their sails on Monenday morn
With all the speed they may;
They have landed in Noroway
Upon a Wednesday.

II *The Return*

The men of Norway seem to have insulted their Scottish guests by hinting that they were staying too long, and this caused Sir Patrick Spens to sail for home without waiting for fair weather.

'Make ready, make ready, my merry men all!
  Our good ship sails the morn.'—
'Now ever alack, my master dear,
  I fear a deadly storm.

'I saw the new moon late yest'r-e'en
  With the old moon in her arm;
And if we go to sea, master,
  I fear we'll come to harm.'

They had not sail'd a league, a league,
  A league but barely three,
When the lift grew dark, and the wind blew loud,
  And gurly grew the sea.

The anchors brake, and the topmast lap,
  It was such a deadly storm;
And the waves came o'er the broken ship
  Till all her sides were torn.

'O where will I get a good sailor
  To take my helm in hand,
While I go up to the tall topmast
  To see if I can spy land?'—

'O here am I, a sailor good,
  To take the helm in hand,
While you go up to the tall topmast,
  But I fear you'll ne'er spy land.'

He had not gone a step, a step,
  A step but barely one,
When a bolt flew out of our goodly ship,
  And the salt sea it came in.

'Go fetch a web o' the silken cloth,
  Another o' the twine,
And wap them into our ship's side,
  And let not the sea come in.'

They fetch'd a web o' the silken cloth,
  Another o' the twine,
And they wapp'd them round that good ship's side,
  But still the sea came in.

O loth, loth were our good Scots lords
  To wet their cork-heeled shoon;
But long ere all the play was play'd
  They wet their hats aboon.

And many was the feather bed
　That flatter'd on the foam;
And many was the good lord's son
　That never more came home.

O long, long may the ladies sit,
　With their fans into their hand,
Before they see Sir Patrick Spens
　Come sailing to the strand!

And long, long may the maidens sit
　With their gold combs in their hair,
A-waiting for their own dear loves!
　For them they'll see no more.

Half-o'er, half-o'er to Aberdour,
　'Tis fifty fathoms deep;
And there lies good Sir Patrick Spens,
　With the Scots lords at his feet!

Anon.

The Pied Piper of Hamelin

Hamelin Town's in Brunswick,
By famous Hanover city;
The river Weser, deep and wide,
Washes its wall on the southern side;
A pleasanter spot you never spied;
But, when begins my ditty,
Almost five hundred years ago,
To see the townsfolk suffer so
From vermin, was a pity.

Rats!
They fought the dogs, and killed the cats,
And bit the babies in the cradles,
And ate the cheeses out of the vats,
And licked the soup from the cooks' own ladles,
Split open the kegs of salted sprats,
Made nests inside men's Sunday hats,
And even spoiled the women's chats,
By drowning their speaking
With shrieking and squeaking
In fifty different sharps and flats.

At last the people in a body
To the Town Hall came flocking:
''Tis clear,' cried they, 'our Mayor's a noddy;
And as for our Corporation—shocking
To think we buy gowns lined with ermine
For dolts that can't or won't determine
What's best to rid us of our vermin!
You hope, because you're old and obese,
To find in the furry civic robe ease?
Rouse up, Sirs! Give your brains a racking,
To find the remedy we're lacking,
Or, sure as fate, we'll send you packing!'
At this the Mayor and Corporation
Quaked with a mighty consternation.

An hour they sate in council,
At length the Mayor broke silence:
'For a guilder I'd my ermine gown sell;
I wish I were a mile hence!
It's easy to bid one rack one's brain—
I'm sure my poor head aches again
I've scratched it so, and all in vain.
Oh for a trap, a trap, a trap!'
Just as he said this, what should hap
At the chamber door but a gentle tap?
'Bless us,' cried the Mayor, 'what's that?'
(With the Corporation as he sat,
Looking little though wondrous fat;
Nor brighter was his eye, nor moister
Than a too-long-opened oyster,
Save when at noon his paunch grew mutinous
For a plate of turtle green and glutinous)
'Only a scraping of shoes on the mat?
Anything like the sound of a rat
Makes my heart go pit-a-pat!'

'Come in!' —the Mayor cried, looking bigger:
And in did come the strangest figure!
His queer long coat from heel to head
Was half of yellow and half of red;
And he himself was tall and thin,
With sharp blue eyes, each like a pin,
And light loose hair, yet swarthy skin,
No tuft on cheek nor beard on chin,
But lips where smiles went out and in—
There was no guessing his kith and kin!
And nobody could enough admire
The tall man and his quaint attire:
Quoth one: 'It's as my great-grandsire,
Starting up at the Trump of Doom's tone,
Had walked this way from his painted tomb-stone!'

He advanced to the council-table:
And, 'Please your honours,' said he, 'I'm able
By means of a secret charm to draw
All creatures living beneath the sun,
That creep or swim or fly or run,
After me so as you never saw!
And I chiefly use my charm
On creatures that do people harm,
The mole and toad and newt and viper;
And people call me the Pied Piper.'
(And here they noticed round his neck
A scarf of red and yellow stripe,
To match with his coat of the self-same cheque;
And at the scarf's end hung a pipe;
And his fingers, they noticed, were ever straying
As if impatient to be playing
Upon this pipe, as low it dangled
Over his vesture so old-fangled.)
'Yet,' said he, 'poor piper as I am,
In Tartary I freed the Cham,
Last June, from his huge swarms of gnats;
I eased in Asia the Nizam
Of a monstrous brood of vampyre-bats;
And as for what your brain bewilders,
If I can rid your town of rats
Will you give me a thousand guilders?'
'One? fifty thousand!'—was the exclamation
Of the astonished Mayor and Corporation.

Into the street the Piper stept,
Smiling first a little smile,
As if he knew what magic slept
In his quiet pipe the while;
Then, like a musical adept,
To blow the pipe his lips he wrinkled,
And green and blue his sharp eyes twinkled
Like a candle-flame where salt is sprinkled;
And ere three shrill notes the pipe uttered,
You heard as if an army muttered;
And the muttering grew to a grumbling;
And the grumbling grew to a mighty rumbling;
And out of the houses the rats came tumbling.

Great rats, small rats, lean rats, brawny rats,
Brown rats, black rats, grey rats, tawny rats,
Grave old plodders, gay young friskers,
Fathers, mothers, uncles, cousins,
Cocking tails and pricking whiskers,
Families by tens and dozens,
Brothers, sisters, husbands, wives—
Followed the Piper for their lives.
From street to street he piped advancing,
And step for step they followed dancing,
Until they came to the river Weser
Wherein all plunged and perished!
—Save one who, stout as Julius Caesar,
Swam across and lived to carry
(As he, the manuscript he cherished)
To Rat-land home his commentary:
Which was, 'At the first shrill notes of the pipe,
I heard a sound as of scraping tripe,
And putting apples, wondrous ripe,
Into a cider-press's gripe:
And a moving away of pickle-tub boards,
And a leaving ajar of conserve-cupboards,
And a drawing the corks of train-oil flasks,
And a breaking the hoops of butter-casks;
And it seemed as if a voice
(Sweeter far than by harp or by psaltery
Is breathed) called out, Oh rats, rejoice!
The world is grown to one vast dry-saltery!
So, munch on, crunch on, take your nuncheon,
Breakfast, supper, dinner, luncheon!
And just as a bulky sugar-puncheon,
All ready staved, like a great sun shone
Glorious scarce an inch before me,
Just as methought it said, Come, bore me!
—I found the Weser rolling o'er me.'

You should have heard the Hamelin people
Ringing the bells till they rocked the steeple.
'Go,' cried the Mayor, 'and get long poles!
Poke out the nests and block up the holes!
Consult with carpenters and builders,
And leave in our town not even a trace
Of the rats!'—when suddenly, up the face
Of the Piper perked in the market-place,
With a 'First, if you please, my thousand guilders!'

A thousand guilders! The Mayor looked blue;
So did the Corporation too.
For council dinners made rare havoc
With Claret, Moselle, Vin-de-Grave, Hock;
And half the money would replenish
Their cellar's biggest butt with Rhenish.
To pay this sum to a wandering fellow
With a gipsy coat of red and yellow!
'Beside,' quoth the Mayor with a knowing wink,
'Our business was done at the river's brink;
We saw with our eyes the vermin sink,
And what's dead can't come to life, I think.
So, friend, we're not the folks to shrink
From the duty of giving you something to drink,
And a matter of money to put in your poke;
But as for the guilders, what we spoke
Of them, as you very well know, was in joke.
Besides, our losses have made us thrifty.
A thousand guilders! Come, take fifty!'

The Piper's face fell, and he cried,
'No trifling! I can't wait. Beside,
I've promised to visit by dinner time
Bagdad, and accept the prime
Of the Head-Cook's pottage, all he's rich in,
For having left, in the Caliph's kitchen,
Of a nest of scorpions no survivor—
With him I proved no bargain-driver,
With you, don't think I'll bate a stiver!
And folks who put me in a passion
May find me pipe to another fashion.'

'How?' cried the Mayor, 'd'ye think I'll brook
Being worse treated than a Cook?
Insulted by a lazy ribald
With idle pipe and vesture piebald?
You threaten us, fellow? Do your worst,
Blow your pipe there till you burst!'

Once more he stept into the street;
And to his lips again
Laid his long pipe of smooth straight cane;
And ere he blew three notes (such sweet
Soft notes as yet musician's cunning
Never gave the enraptured air)
There was a rustling, that seemed like a bustling
Of merry crowd justling at pitching and hustling,
Small feet were pattering, wooden shoes clattering,
Little hands clapping and little tongues chattering,
And, like fowls in a farm-yard when barley is scattering,
Out came the children running.
All the little boys and girls,
With rosy cheeks and flaxen curls,
And sparkling eyes and teeth like pearls,
Tripping and skipping, ran merrily after
The wonderful music with shouting and laughter.

The Mayor was dumb, and the Council stood
As if they were changed into blocks of wood,
Unable to move a step, or cry
To the children merrily skipping by—
And could only follow with the eye
That joyous crowd at the Piper's back.
But how the Mayor was on the rack,
And the wretched Council's bosoms beat,
As the Piper turned from the High Street
To where the Weser rolled its waters
Right in the way of their sons and daughters!
However he turned from South to West,
And to Koppelberg Hill his steps addressed,
And after him the children pressed;
Great was the joy in every breast.
'He never can cross that mighty top!
He's forced to let the piping drop,
And we shall see our children stop!'

When, lo, as they reached the mountain's side,
A wondrous portal opened wide,
As if a cavern was suddenly hollowed;
And the Piper advanced and the children followed,
And when all were in to the very last,
The door in the mountain-side shut fast.
Did I say, all? No! One was lame,
And could not dance the whole of the way;
And in after years, if you would blame
His sadness, he was used to say,—
'It's dull in our town since my playmates left!
I can't forget that I'm bereft
Of all the pleasant sights they see,
Which the Piper also promised me.
For he led us, he said, to a joyous land,
Joining the town and just at hand,
Where waters gushed and fruit-trees grew,
And flowers put forth a fairer hue,
And everything was strange and new;
The sparrows were brighter than peacocks here,
And their dogs outran our fallow deer,
And honey-bees had lost their stings,
And horses were born with eagles' wings:
And just as I became assured
My lame foot would be speedily cured,
The music stopped and I stood still,
And found myself outside the Hill,
Left alone against my will,
To go now limping as before,
And never hear of that country more!'

Alas, alas for Hamelin!
There came into many a burgher's pate
A text which says, that Heaven's Gate
Opes to the Rich at as easy rate
As the needle's eye takes a camel in!
The Mayor sent East, West, North and South,
To offer the Piper, by word of mouth,
Wherever it was men's lot to find him,
Silver and gold to his heart's content,
If he'd only return the way he went,
And bring the children behind him.

But when they saw ’twas a lost endeavour,
And Piper and dancers were gone for ever,
They made a decree that lawyers never
Should think their records dated duly
If, after the day of the month and year,
These words did not as well appear,
‘And so long after what happened here
On the Twenty-second of July,
Thirteen-hundred and seventy-six’:
And the better in memory to fix
The place of the children’s last retreat,
They called it, the Pied Piper’s Street—
Where any one playing on pipe or tabor
Was sure for the future to lose his labour
Nor suffered they hostelry or tavern
To shock with mirth a street so solemn;
But opposite the place of the cavern
They wrote the story on a column,
And on the great Church-Window painted
The same, to make the world acquainted
How their children were stolen away;
And there it stands to this very day.

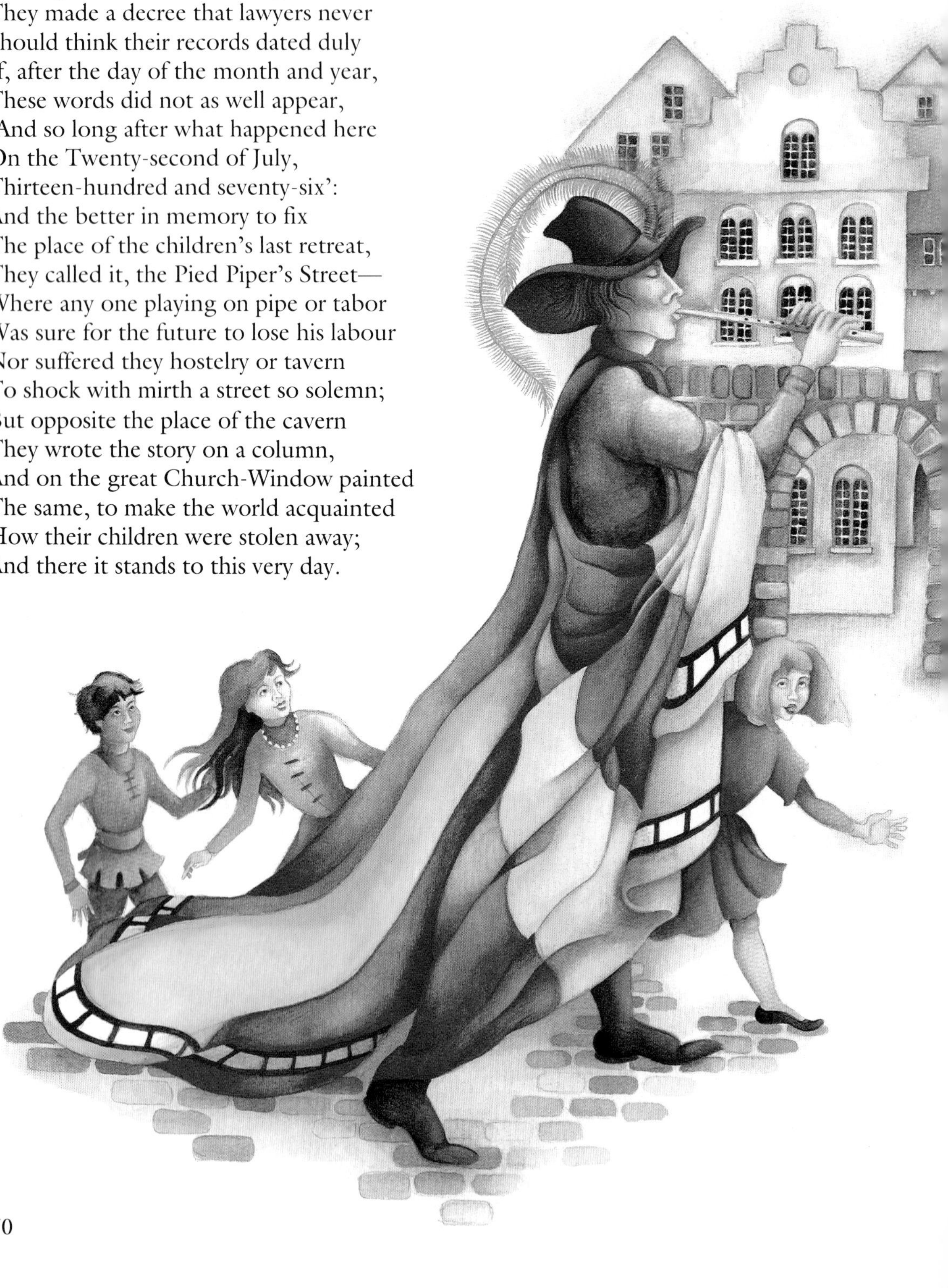

And I must not omit to say
That in Transylvania there's a tribe
Of alien people that ascribe
The outlandish ways and dress
On which their neighbours lay such stress
To their fathers and mothers having risen
Out of some subterraneous prison
Into which they were trepanned
Long time ago in a mighty band
Out of Hamelin town in Brunswick land,
But how or why, they don't understand.

Robert Browning

P. C. Plod versus the Dale St Dogstrangler

For several months
Liverpool was held in the grip of fear
by a dogstrangler most devilish,
who roamed the streets after dark
looking for strays. Finding one
he would tickle it seductively
about the body to gain its confidence,
then lead it down a deserted backstreet
where he would strangle the poor brute.
Hardly a night passed without somebody's
faithful fourlegged friend being dispatched
to that Golden Kennel in the sky.

The public were warned:
At the very first sign
of anything suspicious,
ring Canine-nine-nine.

Nine o'clock on the evening of January 11th
sees P. C. Plod on the corner
of Dale St and Sir Thomas St
disguised as a Welsh collie.
It is part of a daring plan to apprehend the strangler.
For though it is a wet and moonless night,
Plod is cheered in the knowledge
that the whole of the Liverpool City Constabulary
is on the beat that night disguised as dogs.

Not ten minutes earlier, a pekinese
(Policewoman Hodges)
had scampered past on her way to Clayton Square.

For Plod, the night passed uneventfully
and so in the morning he was horrified to learn
that no less than fourteen policemen and policewomen
had been tickled and strangled during the night.

The public were horrified
The Commissioner aghast
Something had to be done
And fast.

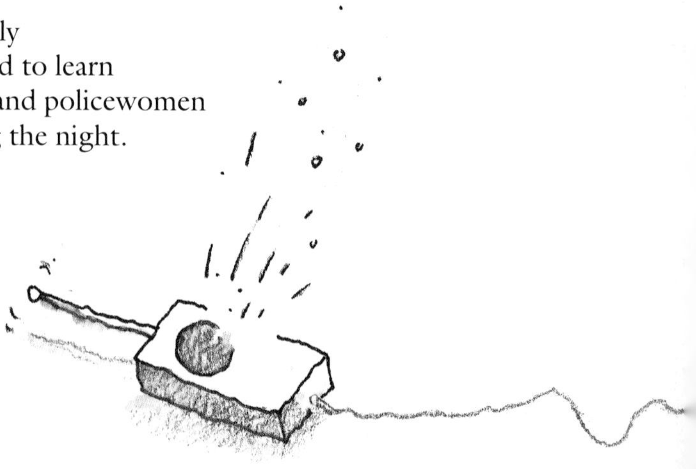

P. C. Plod (wise as a brace of owls)
met the challenge magnificently
and submitted an idea so startling in its vision
so audacious in its conception
that the Commissioner gasped
before ordering all dogs in the city
to be thereinafter disguised as fuzz.
The plan worked
and the dogstrangler was heard of no more.

Cops and mongrels
like P. C.s in a pod
To a grateful public
Plod was God.

So next time you're up in Liverpool
take a closer look
at that policeman on pointduty,
he might well be a copper spaniel.

Roger McGough

Uncle Alfred's Long Jump

When Mary Rand
Won the Olympic Long Jump,
My Auntie Hilda
Paced out the distance
On the pavement outside her house.
'Look at that!'
She shouted challengingly
At the dustman, the milkman, the grocer,
Two Jehovah's Witnesses
And a male St. Bernard
Who happened to be passing.
'A girl, a girl did that;
If you men are so clever
Let's see what you can do.'
Nobody took up the challenge
Until Uncle Alfred trudged home
Tired from the office
Asking for his tea.
'Our Mary did that!'
Said Auntie Hilda proudly
Pointing from the lamppost
To the rose-bush by her gate.
'You men are so clever,
Let's see how near
That rose-bush you end up.'

His honour and manhood at stake,
Uncle Alfred put down his bowler
His brief-case and his brolly
And launched himself
Into a fifty yard run-up.
'End up at that rose-bush,'
He puffed mockingly,
'I'll show you where I'll end up.'
His take off from the lamppost
Was a thing of beauty,
But where he ended up
Was in The Royal Infirmary
With both legs in plaster.
'Some kind of record!'
He said proudly to the bone specialist;
While through long nights
In a ward full of coughs and snoring
He dreamed about the washing line
And of how to improve
His high jump technique.

Gareth Owen

Mr Tom Narrow

A scandalous man
Was Mr Tom Narrow,
He pushed his grandmother
Round in a barrow.
And he called out loud
As he rang his bell,
'Grannies to sell!
Old grannies to sell!'

The neighbours said,
As they passed them by,
'This poor old lady
We will not buy.
He surely must be
A mischievous man
To try for to sell
His own dear Gran.'

'Besides,' said another,
'If you ask me,
She'd be very small use
That I can see.'
'You're right,' said a third,
'And no mistake—
A very poor bargain
She'd surely make.'

So Mr Tom Narrow
He scratched his head,
And he sent his grandmother
Back to bed;
And he rang his bell
Through all the town
Till he sold his barrow
For half a crown.

James Reeves

Bishop Hatto

The summer and the autumn had been so wet
That in winter the corn was growing yet;
'Twas a piteous sight to see all around
The grain lie rotting on the ground.

Every day the starving poor
Crowded around Bishop Hatto's door,
For he had a plentiful last-year's store,
And all the neighbourhood could tell
His granaries were furnish'd well.

At last Bishop Hatto appointed a day
To quiet the poor without delay;
He bade them to his great barn repair,
And they should have food for the winter there.

Rejoiced such tidings good to hear,
The poor folk flock'd from far and near;
The great barn was full as it could hold
Of women and children, and young and old.

Then when he saw it could hold no more,
Bishop Hatto he made fast the door,
And while for mercy on Christ they call,
He set fire to the barn and burnt them all.

'I' faith, 'tis an excellent bonfire!' quoth he,
'And the country is greatly obliged to me,
For ridding it in these times forlorn
Of rats, that only consume the corn.'

So then to his palace returned he,
And he sat down to supper merrily,
And he slept that night like an innocent man.
But Bishop Hatto never slept again.

In the morning as he enter'd the hall,
Where his picture hung against the wall,
A sweat like death all over him came;
For the rats had eaten it out of the frame.

As he look'd there came a man from his farm,
He had a countenance white with alarm;
'My lord, I open'd your granaries this morn,
And the rats had eaten all your corn.'

Another came running presently,
And he was pale as pale could be;
'Fly! My Lord Bishop, fly,' quoth he,
'Ten thousand rats are coming this way—
The Lord forgive you for yesterday!'

'I'll go to my tower on the Rhine,' replied he;
''Tis the safest place in Germany;
The walls are high, and the shores are steep,
And the stream is strong, and the water deep.'

Bishop Hatto fearfully hasten'd away,
And he cross'd the Rhine without delay,
And reach'd his tower, and barr'd with care
All the windows, doors, and loopholes there.

He laid him down and closed his eyes,
But soon a scream made him arise;
He started, and saw two eyes of flame
On his pillow from whence the screaming came.

He listen'd and look'd; it was only the cat;
But the Bishop he grew more fearful for that,
For she sat screaming, mad with fear,
At the army of rats that was drawing near.

For they have swum over the river so deep,
And they have climb'd the shores so steep,
And up the tower their way is bent,
To do the work for which they were sent.

They are not to be told by the dozen or score;
By thousands they come, and by myriads and more;
Such numbers had never been heard of before,
Such a judgement had never been witness'd of yore.

Down on his knees the Bishop fell,
And faster and faster his beads did he tell,
As louder and louder drawing near
The gnawing of their teeth he could hear.

And in at the windows, and in at the door,
And through the walls helter-skelter they pour,
And down from the ceiling, and up through the floor,
From the right and the left, from behind and before,
From within and without, from above and below,
And all at once to the Bishop they go.

They have whetted their teeth against the stones,
And now they pick the Bishop's bones;
They gnaw'd the flesh from every limb,
For they were sent to do judgement on him!

Robert Southey

Count Carrots

(from a Bohemian folk-tale called *Rübezahl*)

A small wind lightly
steps over the harebells . . .

Like tall ragged kings
rise the fir trees of Bohemia . . .

And I remember too
the scarlet and purple berries . . .

He's the giant of the mountains;
they call him Count Carrots.
How he hates that nickname.
Let me tell you how he came by it.

Well—there was that princess
who—Persephone-like—
had strayed from her companions.
Perhaps you know the story.

It was all meadows and summer.
It was harebells and clover.
It was tall marguerites.
It was field flowers thousands and thousands,

or so it seemed.
The princess
was tempted to pick the best posy of all.
She ran this way and that, further and further away.

And the voices behind her grew fainter,
and the sky above her grew bluer,
and the sweet meadows engulfed her.
And suddenly—vast arms lifted her into the air.

The princess screamed. Or she may have fainted.
Then the giant of the mountain lifted her onto his shoulders.
He was swarthy and hairy; he was gnarled and muscled like trees.
His stride was long. The princess vanished from sight.

What had her companions been doing?
Asking the daisies who loved them.
Putting buttercups under each other's chin.
'If it shows gold in reflection, it means you like butter.'

Only later they missed her.
The consternation of it;
the runnings to and fro;
the calling, over and over.

'What shall we tell them at home?'
And who will comfort ever
the queen in tears,
the king in despair?

There was one other who heard the news
of the disappearance. He was the prince
whom the princess loved.
He saddled his horse and set out in search of her.

As for the giant, he carried the princess
to his cave under the mountains.
Some say he brought her gifts of precious stones
to tempt her to love him. This is untrue,

he was simpler than that. He brought her, I think,
bilberries from the forest, baskets of raspberries,
mushrooms, many sorts, which Bohemia excels in,
and clumsy importunings, day after day.

He brought her wild strawberries, gathered from steep hills.
She was used to sugar and cream; she was used to pretty bowls
from which to eat them. He roasted venison: the smoke
stung her eyes, she said. She feared the spluttering fat.

The fact is—to paddle your feet in a mountain stream,
shallow and fast and cold as molten ice, water which rushes and swirls
over white pebbles,—to paddle your feet in this on a hot day
is pleasant and delightful: to wash in it, day after day,

indubitably cold. So the princess had discovered.
Besides, she missed her companions, she missed the court and the fun.
She missed, of course, her mother and father, she said.
She missed her little dog, Peep. And she missed her prince.

'Ah, giant, you brought me here against my inclinations.
I am not made for this rocky existence.
Forests are well enough for Sunday hikes.
My dog, Peep, would enjoy them.' Here her tears rolled down.

The giant, slow and ponderous, then had an idea,
which he should have thought of before.
He had some magic. He had a field of carrots.
He brought some to the princess. 'These carrots,' he said,

'can be changed, as you will, by magic into whomever,
whatever, you wish, say, your dog, Peep.'
So the princess wished, and one of the carrots
became her little dog. There he stood, yapping.

The princess, laughing in pleasure and disbelief,
stroked him and patted him and took him into her arms,
then put him down again; the little carrot dog
wagged his tail and sniffed at the venison. He was as like as like.

Then the princess went to work and said to one carrot:
'You'll be friend Sylvia.' And there Sylvia stood,
and laughed and embraced her, and was no carrot at all.
And so: 'You shall be Alice, you shall be John.'

Then the carrots were turned into friends and footmen,
chambermaids, courtiers, horses to ride on, by the princess,—
even goblets to drink from. Ah, but she had a good time.
She was gracious to the giant. She didn't see much of him.

She had a proper court. It was almost like home.
They went riding at dawn. In the evening they danced.
Then on the third morning, or was it the fourth?
her horse as she rode him began somewhat to droop.

A fine chestnut he was. What could be up with him?
He could hardly manage it back to the caves;
his steps stumbling, his very flesh shrunken.
The princess was anxious and looked around for help.

'Sylvia,' she cried to her friend, the groom not being around.
But Sylvia too looked pale and complained of her head.
Something was wrong with Sylvia . . . Something was very wrong!
The groom she had called for lay in a ditch, dead,

and suddenly turned back into a carrot again.
Yes, carrots shrivel when out of the ground, and so
one by one as the carrots died so did her friends
shrivel and turn into carrots again—the magic gone.

That night at dinner only one servant was left to
pour out her wine. As he poured, he bent at the waist
more than he should. He drooped and tottered off;
and the stem of the wine-glass bent too. The wine spilt red.

Her face full of shock and woe, the princess went to the giant,
who looked guilty and sorry. But then he said:
'Darling princess, the fields are full of carrots.
I can bring you carrots freshly each day, to replace those lost.'

Well, so he did, but the princess felt uneasy;
until she hit on a cunning plan. 'Giant,' she pleaded,
'I fear you may one day run out of carrots. Count them for me
to see how many there are.'—So he did: one by one.

That would take him a day and a bit. His back was turned,
and he bent over the furrows with furrowed brow.
Then the princess picked some carrots, the freshest, the strongest,
and turned two into horses, and one into the prince,

the semblance of him whom she loved.
They rode away with the speed of a wish,
through forests of pine,
through thickets, past mountain streams,

into the valley below.
(Do not fear, do not falter,
do not yet fall behind.
Good Hope, stay by my side,)

so the princess prayed.
And they are fortunate
whom a vision of love
accompanies.

Meanwhile what of the giant?
Ah, but that booby
was still counting his carrots:
'Five-hundred . . . six-hundred . . .
   six-hundred-and-seventy-two . . .

Did he try to pursue her?
The story says he did.
Surely, he lumbered one day out of the forest,
to knock her up, knock her down, rap for some reply. . .

And no reply ever: the castle ears
closed tight to his bellowing lungs; the castle gates
forever shut to him; the curtains too,
though he reached to the topmost storey, drawn to his gaze.

Into the sulky night he plodded like thunder,
and the small pillow, rosy in lamplight, whispered
to the burly wind beating against the door:
'You will never bluster your way into *my* down.'

Henceforth, the giant was called Count Carrots by all;
a nickname he hates, as I told you at the beginning.
Woe to him who so calls him in mischief. Let the
impudent traveller, shouting his name, beware.

When I was small, I called his name to the forest:
'Count Carrots! Count Carrots!' then leapt into bed, half in fear.
He didn't come for me though. Could it be that perhaps he forgave me?
He loves children, they say.—May the forest stay green for him ever.

Gerda Mayer

The Lady of Shalott

Part I

On either side the river lie
Long fields of barley and of rye,
That clothe the wold and meet the sky;
And thro' the field the road runs by
  To many-tower'd Camelot;
And up and down the people go,
Gazing where the lilies blow
Round an island there below,
  The island of Shalott.

Willows whiten, aspens quiver,
Little breezes dusk and shiver
Thro' the wave that runs for ever
By the island in the river
  Flowing down to Camelot.
Four gray walls, and four gray towers,
Overlook a space of flowers,
And the silent isle imbowers
  The Lady of Shalott.

By the margin, willow-veil'd,
Slide the heavy barges trail'd
By slow horses; and unhail'd
The shallop flitteth silken-sail'd
  Skimming down to Camelot:
But who hath seen her wave her hand?
Or at the casement seen her stand?
Or is she known in all the land,
  The Lady of Shalott?

Only reapers, reaping early
In among the bearded barley,
Hear a song that echoes cheerly
From the river winding clearly,
  Down to tower'd Camelot;
And by the moon the reaper weary,
Piling sheaves in uplands airy,
Listening, whispers ''Tis the fairy
  Lady of Shalott.'

Part II

There she weaves by night and day
A magic web with colours gay.
She has heard a whisper say,
A curse is on her if she stay
  To look down to Camelot.
She knows not what the curse may be,
And so she weaveth steadily,
And little other care hath she,
  The Lady of Shalott.

And moving thro' a mirror clear
That hangs before her all the year,
Shadows of the world appear.
There she sees the highway near
  Winding down to Camelot:
There the river eddy whirls,
And there the surly village-churls,
And the red cloaks of market girls,
  Pass onward from Shalott.

Sometimes a troop of damsels glad,
An abbot on an ambling pad,
Sometimes a curly shepherd-lad,
Or long-hair'd page in crimson clad,
    Goes by to tower'd Camelot;
And sometimes thro' the mirror blue
The knights come riding two and two:
She hath no loyal knight and true,
    The Lady of Shalott.

But in her web she still delights
To weave the mirror's magic sights,
For often thro' the silent nights
A funeral, with plumes and lights,
    And music, went to Camelot:
Or when the moon was overhead,
Came two young lovers lately wed;
'I am half sick of shadows,' said
    The Lady of Shalott.

Part III

A bow-shot from her bower-eaves,
He rode between the barley-sheaves,
The sun came dazzling thro' the leaves,
And flamed upon the brazen greaves
  Of bold Sir Lancelot.
A red-cross knight for ever kneel'd
To a lady in his shield,
That sparkled on the yellow field,
  Beside remote Shalott.

The gemmy bridle glitter'd free,
Like to some branch of stars we see
Hung in the golden Galaxy.
The bridle bells rang merrily
  As he rode down to Camelot:
And from his blazon'd baldric slung
A mighty silver bugle hung,
And as he rode his armour rung,
  Beside remote Shalott.

All in the blue unclouded weather
Thick-jewell'd shone the saddle-leather,
The helmet and the helmet-feather
Burn'd like one burning flame together,
  As he rode down to Camelot.
As often thro' the purple night,
Below the starry clusters bright,
Some bearded meteor, trailing light,
  Moves over still Shalott.

His broad clear brow in sunlight glow'd;
On burnish'd hooves his war-horse trode;
From underneath his helmet flow'd
His coal-black curls as on he rode,
  As he rode down to Camelot.
From the bank and from the river
He flash'd into the crystal mirror,
'Tirra lirra,' by the river
  Sang Sir Lancelot.

She left the web, she left the loom,
She made three paces thro' the room,
She saw the water-lily bloom,
She saw the helmet and the plume,
    She look'd down to Camelot.
Out flew the web and floated wide;
The mirror crack'd from side to side;
'The curse is come upon me!' cried
    The Lady of Shalott.

Part IV

In the stormy east-wind straining,
The pale yellow woods were waning,
The broad stream in his banks complaining,
Heavily the low sky raining
  Over tower'd Camelot;
Down she came and found a boat
Beneath the willow left afloat,
And round about the prow she wrote
  The Lady of Shalott.

And down the river's dim expanse—
Like some bold seer in a trance,
Seeing all his own mischance—
With a glassy countenance
    Did she look to Camelot.
And at the closing of the day,
She loosed the chain, and down she lay;
The broad stream bore her far away,
    The Lady of Shalott.

Lying, robed in snowy white
That loosely flew to left and right—
The leaves upon her falling light—
Thro' the noises of the night
    She floated down to Camelot:
And as the boat-head wound along
The willowy hills and fields among,
They heard her singing her last song,
    The Lady of Shalott.

Heard a carol, mournful, holy,
Chanted loudly, chanted lowly,
Till her blood was frozen slowly,
And her eyes were darken'd wholly,
    Turn'd to tower'd Camelot;
For ere she reach'd upon the tide
The first house by the water-side,
Singing in her song she died,
    The Lady of Shalott.

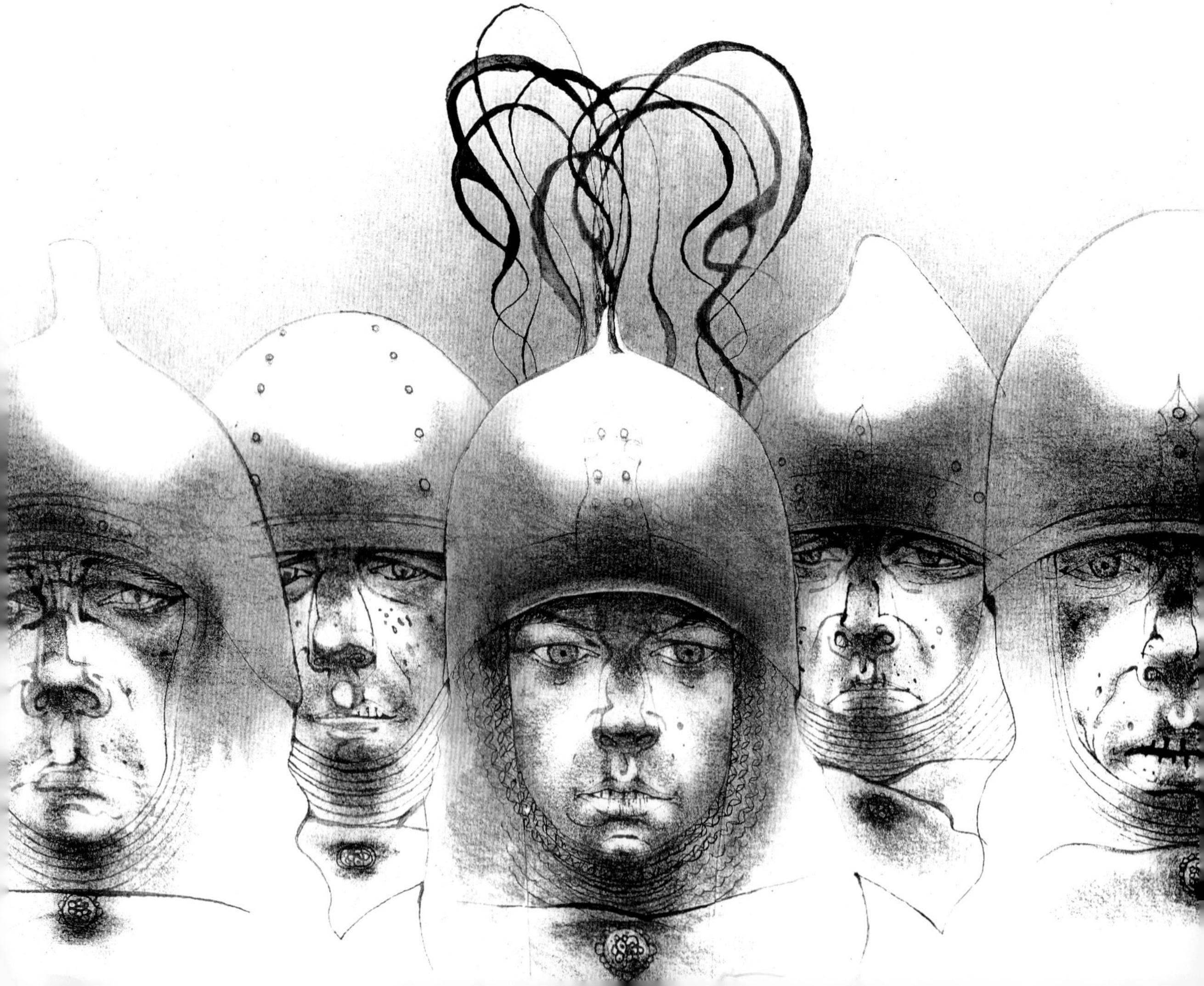

Under tower and balcony,
By garden wall and gallery,
A gleaming shape she floated by,
Dead-pale between the houses high,
Silent into Camelot.
Out upon the wharfs they came,
Knight and burgher, lord and dame,
And round the prow they read her name,
The Lady of Shalott.

Who is this? and what is here?
And in the lighted palace near
Died the sound of royal cheer;
And they cross'd themselves for fear,
All the Knights at Camelot:
But Lancelot mused a little space;
He said, 'She has a lovely face;
God in His mercy lend her grace,
The Lady of Shalott.'

Lord Tennyson

The Apple-Tree Man

The farmer sleeps under a printed stone.
The farm it fell to the Youngest Son.
The Eldest has neither a hearth nor home.
A sorrow and shame, said the Apple-Tree Man.

The Youngest he lends him an orchard green,
An ox and an ass and a handful of grain
And their Grannie's old cottage in Watery Lane.
It's not what it sounds, said the Apple-Tree Man.

For the grain it was mouldy, the roof it had flown,
And never a fruit had the apple-trees grown,
The dunk was all skin and the ox was all bone.
Here's a how-do-ye-do, said the Apple-Tree Man.

The Eldest he neither did mutter nor moan.
He found all the slates and he nailed them back on.
He laid the grass low that was lanky and long.
Will-o'-the-work! said the Apple-Tree Man.

He cured him his beasts with the words of a charm.
The ox and the ass to the orchard are gone,
And the apple-trees flourish as never they've done.
Sun's coming up, said the Apple-Tree Man.

Said Youngest to Eldest, 'Now pray understand
When Quarter Day comes you must pay on demand
And dap down the rent on the palm of my hand.'
Brotherly love! said the Apple-Tree Man.

But the Eldest had hardly the price of a pin.
Though he worked and he worried his profit was thin.
'It's wrecked and it's ruined,' he said, 'that I am!'
Can't have that, said the Apple-Tree Man.

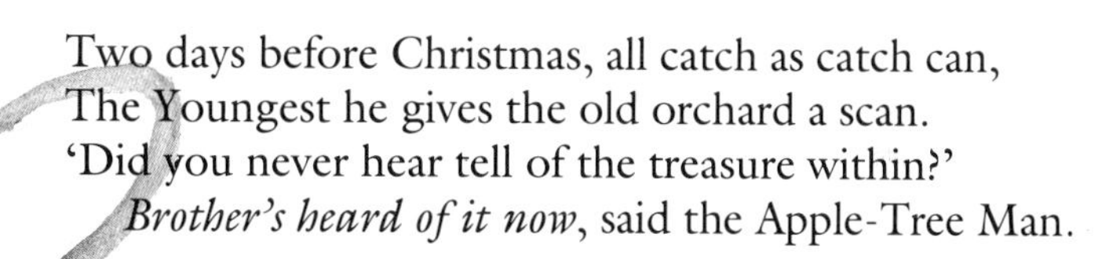

Two days before Christmas, all catch as catch can,
The Youngest he gives the old orchard a scan.
'Did you never hear tell of the treasure within?'
Brother's heard of it now, said the Apple-Tree Man.

'Day after tomorrow when midnight is come
And the beasts in the shippen no longer are dumb
To task 'em and ask 'em,' he said, 'is my plan.'
They'll tell you no lies, said the Apple-Tree Man.

'So brother,' the Youngest said, 'wake me betimes
Before Christmas comes and the church clock it chimes
And a sixpence I'll slice off your rent for a span.'
Such bounteousness! said the Apple-Tree Man.

But the Eldest hangs holly all up in a chain
And he gives to the ox the sweet hay and good grain
And to the old donkey he gives just the same.
Wassail! Wassail! said the Apple-Tree Man.

Then the Eldest his cider mug fills to the brim
And gives to his apple-trees out in the dim.
'I wish you Good Christmas,' he said, 'where you stand!'
Look under my roots, said the Apple-Tree Man.

Says the Eldest, 'I'm blest, but it's magic that's planned
For the earth and the stones are all softer than sand.'
And a chest full of gold he digs up with his hands.
Bide quiet and hide it, says the Apple-Tree Man.

At midnight the Eldest calls Youngest to rise
And down he comes running, the sleep in his eyes.
'Dear ox and dear donkey, please tell if you can
Where lies the gold treasure that's under my land?'

The gold and the treasure are taken and gone
And you never shall find it by moon or by sun
Though all the wide world you may search and may scan,
Said the ox and the ass and the Apple-Tree Man.

Charles Causley

Lost

In a terrible fog I once lost my way,
Where I had wandered I could not say,
I found a signpost just by a fence,
But I could not read it, the fog was so dense.
Slowly but surely, frightened to roam,
I climbed up the post for my nearest way home,
Striking a match I turned cold and faint,
These were the words on it, 'Mind the wet paint.'

James Godden

The Sad Story of a Little Boy That Cried

Once a little boy, Jack, was, oh! ever so good,
Till he took a strange notion to cry all he could.

So he cried all the day, and he cried all the night,
He cried in the morning, and in the twilight;

He cried till his voice was as hoarse as a crow,
And his mouth grew so large it looked like a great O.

It grew at the bottom, and grew at the top;
It grew till they thought that it never would stop.

Each day his great mouth grew taller and taller,
And his dear little self grew smaller and smaller.

At last, that same mouth grew so big that—alack!—
It was only a mouth with a border of Jack.

Anon.

Annabell and the Witches

Once upon a time there was a girl called Annabell.
Her mother said she was—
  a sweet little girl
  a dear little girl
  a nice little girl.
But Annabell didn't like that one little bit,
because Annabell wanted to be
A WITCH!

Not any old kind of witch—
no, Annabell wanted to be
  a nasty witch
  an ugly witch
  a nasty, ugly, evil witch . . .

The only trouble was
Annabell didn't know any witches,
so she was having a hard time
learning how to be nasty, ugly and evil.

Until, one day, she had a brilliant idea:
She looked up 'Witch' in the phone book!
And she found:
WITCH, N., U. & E. Ltd, On the Common. And
  then a phone number: Nasty Weather 1200.
But when she rang the number
all she got was
a nasty, ugly, evil slurping noise—
the witches' phone had been
                    Cut Off!
So Annabell wrote the address
on the back of her hand,
waited until the weather was *really* nasty
and went to the Common
                    All Alone . . . !

When she got there, she saw
three witches dancing round a big black pot.
They were singing:

> 'Hubble, bubble, boils and spots,
> Gravy full of slimy clots,
> Smelly socks and filthy rags
> And things that live in dustbin bags . . .'

'That sounds *great*!'
Shouted Annabell. 'Let me have a go—
I want to be a witch, too!'

The nastiest, ugliest,
most evil-looking witch spun round.
She glared at Annabell.
Then she asked a nasty question:
'Can you spell?'
'Yes,' said Annabell,
'. . . a bit.'

'You can't be a witch without a spell,'
sang all the witches together.
'And you can't make a spell until you can
S-P-E-L-L!'
'I'll show you,' said
the nastiest, ugliest,
most evil witch.
And she began to sing
in a voice so scratched and tuneless
that it made Annabell's teeth cringe:

'*S* low and slimy, cold and clammy
L ike a lump of
U ncooked liver.
G reen revolting slithering slug.
Slug with silver slime trail oozing,
S-L-U-G I spell and
Slug you be!'

And Annabell felt her body shrink
—smaller and smaller
and her skin become cold and wet.
And suddenly
all she could see
stretching high above her
were stalks of grass
—as tall and thick as apple trees.
And the tiny grains of soil
felt like great rocks
beneath her one wet foot.
And all she could hear was the
'Arkh, Arkh, Arkh!' cackle
of the witches laughing at her.

Then the ugliest, nastiest,
most evil witch stopped laughing.
She cleared her throat
and began to sing another song:

'*P* orker
I n the pigsty
G runting—

Swell my swill-pig fat with lard.
Balloon of bacon, pink and pudgy—
P-I-G I spell and
Pig you be!'

And as the witch sang the last line
of the spell,
Annabell turned back from a slug
into a girl . . .
but only for a moment.
Annabell felt herself swell.
And the pretty pink bow
her mother had tied
at the back of her dress became
a pink curly tail.

And the pretty pink bow
her mother had tied
in her hair turned into
two, huge flapping ears.

'Stop it!' she shouted.
'It's horrible!'
But the only sound
that came out
of her little piggy snout was
'Horkh! Horkh! Horkh!'

Annabell looked up
with her little pink, poggy eyes
and saw the witches
nearly splitting their broomsticks laughing.
(Even the wicked black cat was laughing!)

Then the nastiest, ugliest,
most evil witch
wiped the tears from her eyes
with the filthiest handkerchief
that Annabell had ever seen
and sang:

'*A* s you were—
N ot grunting pig
N or slithering slug but
A s
B efore.
E verything be
L ike it was, and
L et you be
A-N-N-A-B-E-L-L once more.'

And Annabell was back on the Common.
As a girl.
Alone!
The witches had . . . Vanished!
(But she could still hear
a nasty, ugly,
evil laugh
echoing, echoing away in the darkness . . .)

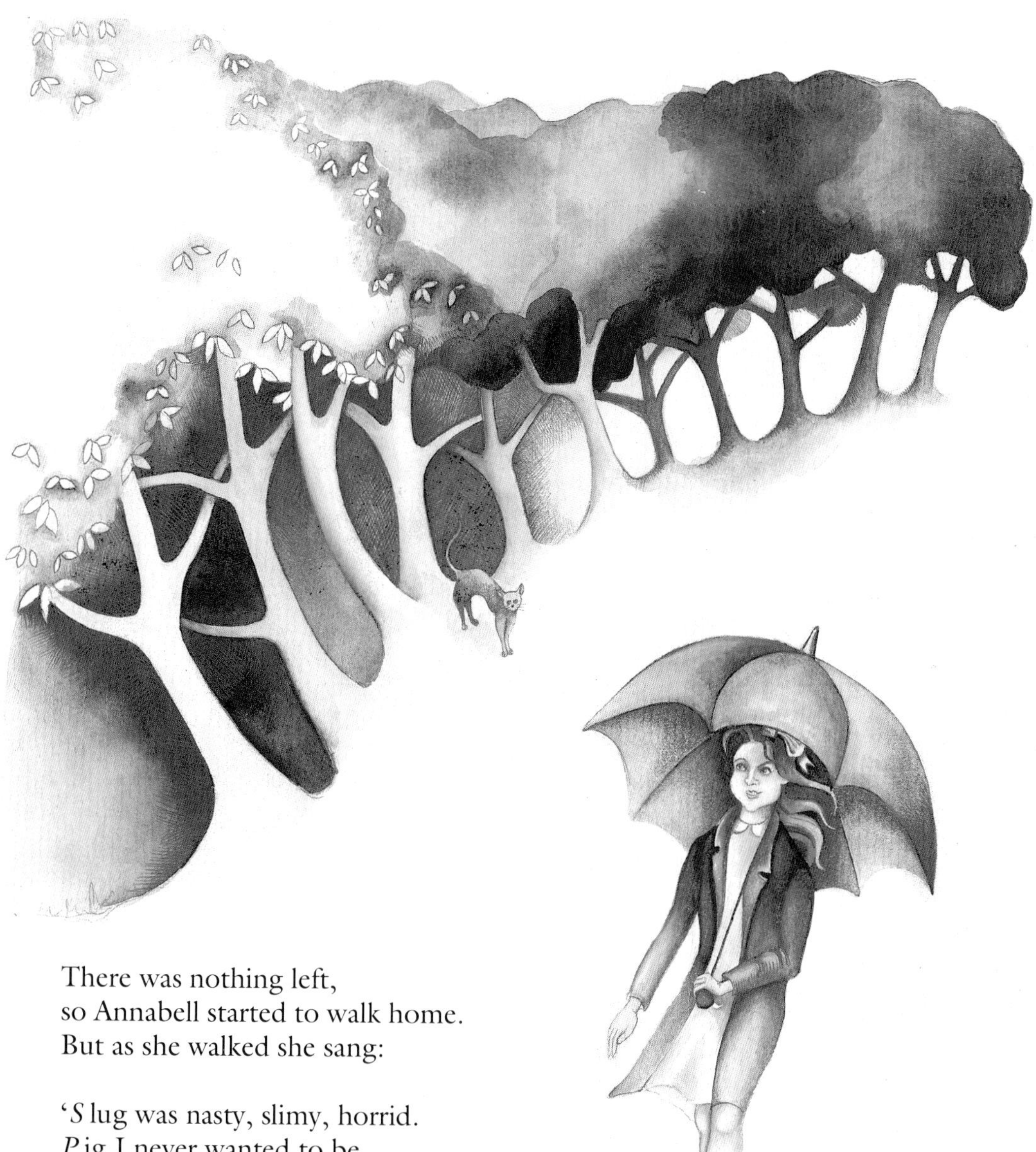

There was nothing left,
so Annabell started to walk home.
But as she walked she sang:

'*S* lug was nasty, slimy, horrid.
P ig I never wanted to be.
E veryone, please
L eave me alone! Just
L et me be Annabell . . .
let me be ME!'

Which wasn't bad
for a beginner . . .
Was it?

Mick Gowar

By St Thomas Water

By St Thomas Water
Where the river is thin
We looked for a jam-jar
To catch the quick fish in.
Through St Thomas Church-yard
Jessie and I ran
The day we took the jam-pot
Off the dead man.

On the scuffed tombstone
The grey flowers fell,
Cracked was the water,
Silent the shell.
The snake for an emblem
Swirled on the slab,
Across the beach of sky the sun
Crawled like a crab.

'If we walk,' said Jessie,
'Seven times round,
We shall hear a dead man
Speaking underground.'
Round the stone we danced, we sang,
Watched the sun drop,
Laid our heads and listened
At the tomb-top.

Soft as the thunder
At the storm's start
I heard a voice as clear as blood,
Strong as the heart.
But what words were spoken
I can never say,
I shut my fingers round my head,
Drove them away.

'What are those letters, Jessie,
Cut so sharp and trim
All round this holy stone
With earth up to the brim?'
Jessie traced the letters
Black as coffin-lead.
'*He is not dead but sleeping*,'
Slowly she said.

I looked at Jessie,
Jessie looked at me,
And our eyes in wonder
Grew wide as the sea.
Past the green and bending stones
We fled hand in hand,
Silent through the tongues of grass
To the river strand.

By the creaking cypress
We moved as soft as smoke
For fear all the people
Underneath awoke.
Over all the sleepers
We darted light as snow
In case they opened up their eyes,
Called us from below.

Many a day has faltered
Into many a year
Since the dead awoke and spoke
And we would not hear.
Waiting in the cold grass
Under a crinkled bough,
Quiet stone, cautious stone,
What do you tell me now?

Charles Causley

Good Taste

Travelling, a man met a tiger, so . . .
He ran. The tiger ran after him
Thinking: How fast I run . . . But

The road thought: How long I am . . . Then
They came to a cliff, yes, the man
Grabbed at an ash root and swung down

Over its edge. Above his knuckles, the tiger.
At the foot of the cliff, its mate. Two mice,
One black, one white, began to gnaw the root.

And by the traveller's head grew one
Juicy strawberry, so . . . hugging the root
The man reached out and plucked the fruit.

How sweet it tasted!

Christopher Logue

Humpty Dumpty's Recitation

In winter, when the fields are white,
I sing this song for your delight—

In spring, when woods are getting green,
I'll try and tell you what I mean.

In summer, when the days are long,
Perhaps you'll understand the song.

In autumn, when the leaves are brown,
Take pen and ink and write it down.

I sent a message to the fish:
I told them 'This is what I wish.'

The little fishes of the sea
They sent an answer back to me.

The little fishes' answer was
'We cannot do it, Sir, because—'

I sent to them again to say
'It will be better to obey.'

The fishes answered with a grin,
'Why, what a temper you are in!'

I told them once, I told them twice:
They would not listen to advice.

I took a kettle large and new,
Fit for the deed I had to do.

My heart went hop, my heart went thump;
I filled the kettle at the pump.

Then someone came to me and said
'The little fishes are in bed.'

I said to him, I said it plain,
'Then you must wake them up again.'

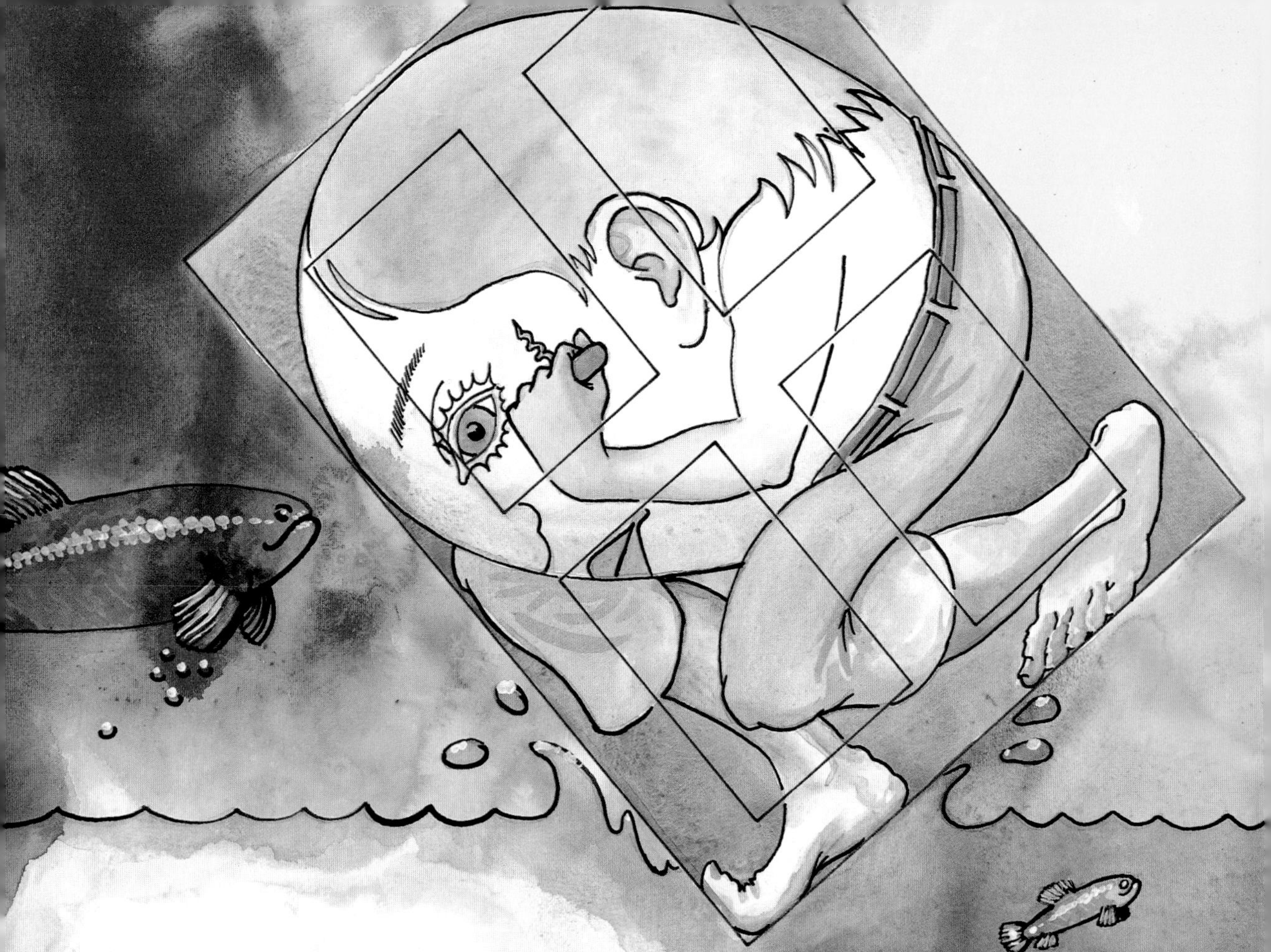

I said it loud and very clear;
I went and shouted in his ear.

But he was very stiff and proud;
He said 'You needn't shout so loud!'

And he was very proud and stiff;
He said 'I'll go and wake them, if—'

I took a corkscrew from the shelf:
I went to wake them up myself.

And when I found the door was locked,
I pulled and pushed and kicked and knocked.

And when I found the door was shut,
I tried to turn the handle, but—

Lewis Carroll

The Owl and the Pussy-Cat

The Owl and the Pussy-cat went to sea
   In a beautiful pea-green boat,
They took some honey, and plenty of money,
   Wrapped up in a five-pound note.
The Owl looked up to the stars above,
   And sang to a small guitar,
'O lovely Pussy! O Pussy, my love,
   What a beautiful Pussy you are,
      You are,
      You are!
   What a beautiful Pussy you are!'

Pussy said to the Owl, 'You elegant fowl!
   How charmingly sweet you sing!
O let us be married! too long we have tarried:
   But what shall we do for a ring?'
They sailed away, for a year and a day,
   To the land where the Bong-tree grows,
And there in a wood a Piggy-wig stood,
   With a ring at the end of his nose,
      His nose,
      His nose,
   With a ring at the end of his nose.

'Dear Pig, are you willing to sell for one shilling
    Your ring?' Said the Piggy, 'I will.'
So they took it away, and were married next day
    By the Turkey who lives on the hill.
They dined on mince, and slices of quince,
    Which they ate with a runcible spoon;
And hand in hand, on the edge of the sand,
    They danced by the light of the moon,
        The moon,
        The moon,
    They danced by the light of the moon.

Edward Lear

Jonah and the Whale

Well, to start with
It was dark
So dark
You couldn't see
Your hand in front of your face;
And huge
Huge as an acre of farm-land.
How do I know?
Well, I paced it out
Length and breadth
That's how.
And if you was to shout
You'd hear your own voice resound,
Bouncing along the ridges of its stomach,
Like when you call out
Under a bridge
Or in an empty hall.
Hear anything?
No not much.
Only the normal
Kind of sounds
You'd expect to hear
Inside a whale's stomach;
The sea swishing far away,
Food gurgling, the wind
And suchlike sounds;
Then there was me screaming for help,
But who'd be likely to hear,
Us being miles from
Any shipping lines
And anyway
Supposing someone did hear,
Who'd think of looking inside a whale?
That's not the sort of thing
That people do.
Smell? I'll say there was a smell.
And cold. The wind blew in
Something terrible from the South
Each time he opened his mouth
Or took a swallow of some tit bit.

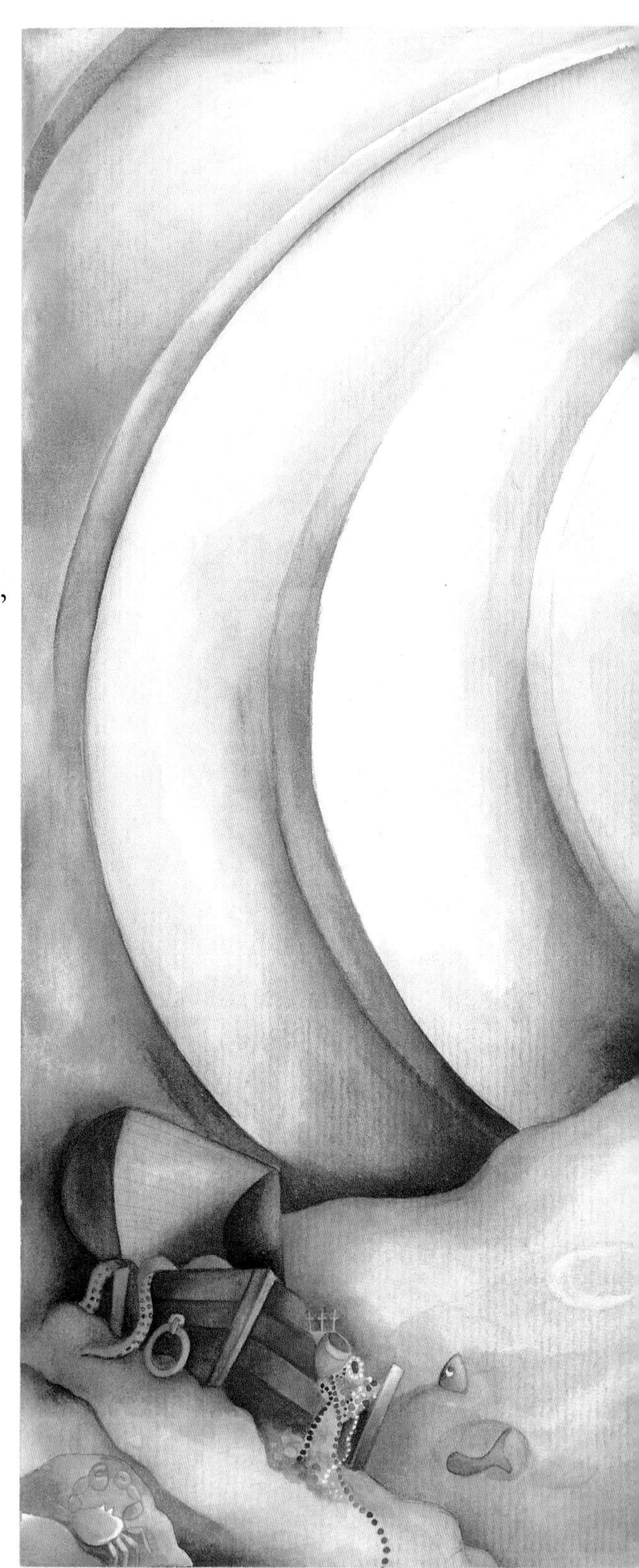

The only way I found
To keep alive at all
Was to wrap my arms
Tight around myself
And race from wall to wall.
Damp? You can say that again;
When the ocean came sluicing in
I had to climb his ribs
To save myself from drowning.
Fibs? You think I'm telling you fibs.
I haven't told the half of it brother.
I'm only giving a modest account
Of what these two eyes have seen
And that's the truth on it.
Here, one thing I'll say
Before I'm done—
Catch me eating fish
From now on.

Gareth Owen

The Jackdaw of Rheims

The Jackdaw sat on the Cardinal's chair!
Bishop and abbot and prior were there;
Many a monk and many a friar,
Many a knight and many a squire,
With a great many more of lesser degree,—
In sooth a goodly company;
And they served the Lord Primate on bended knee.
Never I ween
Was a prouder seen,
Read of in books, or dreamt of in dreams,
Than the Cardinal Lord Archbishop of Rheims!

In and out
Through the motley rout,
That little Jackdaw kept hopping about;
Here and there
Like a dog in a fair,
Over comfits and cates,
And dishes and plates,
Cowl and cope, and rochet and pall,
Mitre and crosier! he hopp'd upon all!
With saucy air,
He perch'd on the chair
Where, in state, the great Lord Cardinal sat
In the great Lord Cardinal's great red hat.
And he peer'd in the face
Of his Lordship's Grace
With a satisfied look, as if he would say,
'We two are the greatest folks here to-day!'
And the priests, with awe,
As such freaks they saw,
Said, 'The Devil must be in that little Jackdaw!'

The feast was over. The board was clear'd.
The flawns and the custards had all disappear'd.
And six little singing-boys,—dear little souls!
In nice clean faces, and nice white stoles,
Came, in order due,
Two by two,
Marching that grand refectory through!

A nice little boy held a golden ewer,
Emboss'd and fill'd with water, as pure
As any that flows between Rheims and Namur,
Which a nice little boy stood ready to catch
In a fine golden hand-basin made to match.
Two nice little boys, rather more grown,
Carried lavender-water, and eau de Cologne;
And a nice little boy had a nice cake of soap,
Worthy of washing the hands of the Pope.
    One little boy more
    A napkin bore
Of the best white diaper, fringed with pink,
And a Cardinal's Hat mark'd in permanent ink.
The great Lord Cardinal turns at the sight
Of these nice little boys dress'd all in white.
    From his finger he draws
    His costly turquoise;
And, not thinking at all about little Jackdaws,
    Deposits it straight
    By the side of his plate,
While the nice little boys on his Eminence wait;
Till, when nobody's dreaming of any such thing,
That little Jackdaw hops off with the ring!

There's a cry and a shout,
And a deuce of a rout,
And nobody seems to know what they're about,
But the monks have their pockets all turn'd inside out.
The friars are kneeling,
And hunting, and feeling
The carpet, the floor, and the walls, and the ceiling.
The Cardinal drew
Off each plum-colour'd shoe,
And left his red stockings exposed to the view;
He peeps, and he feels
In the toes and the heels.
They turn up the dishes. They turn up the plates.
They take up the poker and poke out the grates.
They turn up the rugs.
They examine the mugs.
But, no!—no such thing;—
They can't find THE RING!
And the Abbot declared that, 'when nobody twigg'd it,
Some rascal or other had popp'd in, and prigg'd it!'

The Cardinal rose with a dignified look.
He call'd for his candle, his bell, and his book!

In holy anger and pious grief
He solemnly cursed that rascally thief!
He cursed him at board, he cursed him in bed,
From the sole of his foot to the crown of his head;
He cursed him in sleeping, that every night
He should dream of the devil, and wake in a fright;
He cursed him in eating, he cursed him in drinking,
He cursed him in coughing, in sneezing, in winking;
He cursed him in sitting, in standing, in lying;
He cursed him in walking, in riding, in flying,
He cursed him in living, he cursed him in dying!—
Never was heard such a terrible curse!
But what gave rise
To no little surprise—
Nobody seem'd one penny the worse!

The day was gone.
The night came on.
The Monks and the Friars they search'd till dawn,
When the Sacristan saw,
On crumpled claw,
Come limping a poor little lame Jackdaw!

No longer gay,
As on yesterday;
His feathers all seem'd to be turn'd the wrong way;—
His pinions droop'd—he could hardly stand,—
His head was as bald as the palm of your hand;
His eyes so dim,
So wasted each limb,
That, heedless of grammar, they all cried, 'THAT'S HIM!—
That's the scamp that has done this scandalous thing!
That's the thief that has got my Lord Cardinal's Ring!'

The poor little Jackdaw,
When the Monks he saw,
Feebly gave vent to the ghost of a caw;
And turn'd his bald head, as much as to say,
'Pray, be so good as to walk this way!'
Slower and slower
He limp'd on before,
Till they came to the back of the belfry door,
Where the first thing they saw,
Midst the sticks and the straw,
Was the RING in the nest of that little Jackdaw!

Then the great Lord Cardinal call'd for his book
And off that terrible curse he took.
The mute expression
Served in lieu of confession,
And, being thus coupled with full restitution,
The Jackdaw got plenary absolution!
—When those words were heard,
That poor little bird
Was so changed in a moment, 'twas really absurd.
He grew sleek, and fat.
In addition to that,
A fresh crop of feathers came thick as a mat!

His tail waggled more
Even than before.
But no longer it wagg'd with an impudent air.
No longer he perch'd on the Cardinal's chair.
He hopp'd now about
With a gait devout.

At Matins, at Vespers, he never was out.
And, so far from any more pilfering deeds,
He always seem'd telling the Confessor's beads.
If any one lied,—or if any one swore,—
Or slumber'd in prayer-time and happen'd to snore,
    That good Jackdaw
    Would give a great 'Caw!'
As much as to say, 'Don't do so any more!'
While many remark'd, as his manners they saw,
That they 'never had known such a pious Jackdaw!'
    He long lived the pride
    Of that country-side,
And at last in the odour of sanctity died;
    When, as words were too faint
    His merits to paint,
The Conclave determined to make him a Saint;
And on newly made Saints and Popes, as you know,
It's the custom, at Rome, new names to bestow,
So they canonized him by the name of 'Jim Crow'!

Richard Harris Barham

The Complacent Tortoise

Languid, lethargic, listless and slow,
The tortoise would dally, an image of sloth.
'Immobile!' 'Complacent!' To the hare it was both.

'Enough of your insults, I seek satisfaction.
I'll run you a race and win by a fraction.'
Thus challenged the tortoise one afternoon.
'Right,' said the hare, 'let it be soon.'

They decided they'd race right through the wood
And the tortoise set off as fast as it could.
The hare followed at a leisurely pace
Quite confident it could win the race.

The tortoise thought as it ambled along
'I have never been faster, or quite so strong.'
The hare on the other hand was often inclined
To stop at the roadside and improve its mind.

It read a fable by Aesop deep in the wood
Then of course it set off as fast as it could.
It decided it would put that fable aright
As it sped along with the speed of a light.

Languid, lethargic, listless and slow
The tortoise ran fast as a tortoise could go.
Yet the hare, having decided on saving face,
Quite easily managed to win the race.

'I fccl,' said the tortoise, 'that I've been deceived.
For fables are things I've always believed.
I would love to have won a race so clearly designed
To point out a moral both old and refined.'

'Losing a race would not matter,' the hare said.
'For in speed Mother Nature placed me ahead.
Some fables are things you ought to contest—
Dear tortoise, in mine, I'm afraid you've come last.'

Brian Patten

The Little Turtle

There was a little turtle.
He lived in a box.
He swam in a puddle.
He climbed on the rocks.

He snapped at a mosquito.
He snapped at a flea.
He snapped at a minnow.
And he snapped at me.

He caught the mosquito.
He caught the flea.
He caught the minnow.
But he didn't catch me.

Vachel Lindsay

Merlin and the Snake's Egg

(An early Cornish poem describes how the Druid Merddin,
or Merlin, went early in the morning to seek the magical
snake's egg . . . Robert Graves, *The White Goddess*)

All night the tall young man
    Reads in his book of spells,
Learning the stratagems,
    The chants and diagrams,
The words to serve him well
    When he's the world's magician.
But he needs the snake's egg.

The night is thick as soot,
    The dark wind's at rest,
The fire's low in the grate.
    The black dog stirs and moans
Where he lies at Merlin's feet.
    Dreams trouble his sleep.
Will they search for the snake's egg?

For the purest of magic
    Four things must be found:
Green cress from the river,
    Gold herbs from the ground,
The top twig of the high oak;
    And the snake's round, white egg.
Will they find the snake's egg?

Early, before white dawn
    Disturbs the sleeping world,
Merlin is on his way
    To the forbidden wood.
Glain, the old black dog,
    Walks where his master walked.
They go for the snake's egg.

Glain, are yours the eyes
  To see where the leaf turns?
To know the small, dark hole
  Where the mouse's eye burns?
Will your ears pick up the sound
  Of the mole's breath underground?
Can you find the snake's egg?

Merlin stands at the water's edge,
  At the river's flood.
He stands in the salmon's scales,
  His blood is the salmon's blood.
He swims in the slanting stream,
  In the white foam a whiter gleam.
He has pulled the green cresses.

Merlin stands in the wide field
  Where the small creatures hide.
His long, straight limbs are lost,
  He is changed to a spider.
He crawls on awkward joints, his head
  Moves from side to side.
He has cropped the gold herbs.

Merlin stands beneath the oak.
  Feathers sprout from his arms.
His nose is an owl's hooked nose,
  His voice one of night's alarms,
His eyes are the owl's round eyes.
  Silent and soft he flies.
He has brought down the top twig.

But Glain in the troubled wood
  Steadfastly searches.
The day's last light leans in
  Under the bushes.
And there, like a little moon,
  Pale, round and shining,
He has found the snake's egg.

Leslie Norris

The Unicorn

The Unicorn stood, like a king in a dream,
On the bank of a dark Senegambian stream;
And flaming flamingoes flew over his head,
As the African sun rose in purple and red.

Who knows what the thoughts of a unicorn are
When he shines on the world like a rising star;
When he comes from the magical pages of story
In the pride of his horn and a halo of glory?

He followed the paths where the jungle beasts go,
And he walked with a step that was stately and slow;
But he threw not a shadow and made not a sound,
And his foot was as light as the wind on the ground.

The lion looked up with his terrible eyes,
And growled like the thunder to hide his surprise.
He thought for a while, with a paw in the air;
Then tucked up his tail and turned into his lair.

The gentle giraffe ran away to relate
The news to his tawny and elegant mate,
While the snake slid aside with a venomous hiss,
And the little birds piped: 'There is something amiss!'

But the Unicorn strode with his head in a cloud
And uttered his innocent fancies aloud.
'What a wonderful world!' he was heard to exclaim;
'It is better than books: it is sweeter than fame!'

And he gazed at himself, with a thrill and a quiver,
Reflected in white by the slow-flowing river:
'O speak to me, dark Senegambian stream,
And prove that my beauty is more than a dream!'

He had paused for a word in the midst of his pride,
When a whisper came down through the leaves at his side
From a spying, malevolent imp of an ape
With a twist in his tail and a villainous shape:

'He was made by the stroke of a fanciful pen;
He was wholly invented by ignorant men.
One word in his ear, and one puff of the truth—
And a unicorn fades in the flower of his youth.'

The Unicorn heard, and the demon of doubt
Crept into his heart, and the sun was put out.
He looked in the water, but saw not a gleam
In the slow-flowing deep Senegambian stream.

He turned to the woods, and his shadowy form
Was seen through the trees like the moon in a storm.
And the darkness fell down on the Gambian plain;
And the stars of the Senegal sought him in vain.

He had come like a beautiful melody heard
When the strings of the fiddle are tunefully stirred;
And he passed where the splendours of melody go
When the hand of the fiddler surrenders the bow.

E. V. Rieu

Combinations

A flea flew by a bee. The bee
To flee the flea flew by a fly.
The fly flew high to flee the bee
Who flew to flee the flea who flew
To flee the fly who now flew by.

The bee flew by the fly. The fly
To flee the bee flew by the flea.
The flea flew high to flee the fly
Who flew to flee the bee who flew
To flee the flea who now flew by.

The fly flew by the flea. The flea
To flee the fly flew by the bee.
The bee flew high to flee the flea
Who flew to flee the fly who flew
To flee the bee who now flew by.

The flea flew by the fly. The fly
To flee the flea flew by the bee.
The bee flew high to flee the fly
Who flew to flee the flea who flew
To flee the bee who now flew by.

The fly flew by the bee. The bee
To flee the fly flew by the flea.
The flea flew high to flee the bee
Who flew to flee the fly who flew
To flee the flea who now flew by.

The bee flew by the flea. The flea
To flee the bee flew by the fly.
The fly flew high to flee the flea
Who flew to flee the bee who flew
To flee the fly who now flew by.

Mary Ann Hoberman

The Centipede

A centipede was quite content,
  Until a frog in fun
Said, 'Please, which leg comes after which?'
This worried her to such a pitch,
She lay distracted in the ditch,
  Considering how to run.

Anon.

The Ambitious Ant

The ambitious ant would a-travelling go,
To see the pyramid's wonderful show.
He crossed a brook and a field of rye,
And came to the foot of a haystack high.
'Ah! wonderful pyramid!' then cried he;
'How glad I am that I crossed the sea!'

Amos R. Wells

The Lion and the Echo

The King of the Beasts, deep in the wood,
Roared as loudly as it could.
Right away the echo came back
And the lion thought itself under attack.

'What voice is it that roars like mine?'
The echo replied 'Mine, mine.'

'Who might you be?' asked the furious lion,
'I'm king of this jungle, this jungle is mine.'
And the echo came back a second time,
'This jungle is mine, is mine, is mine.'

The lion swore revenge if only it could
Discover the intruder in the wood.
It roared 'Coward! Come out and show yourself!'
But the fearless echo replied simply '. . . elf.'

'Come out,' roared the lion, 'enough deceit,
Do you fear for your own defeat?'
But all the echo did was repeat
'Defeat . . . defeat . . .'

Frightened by every conceivable sound,
The exhausted lion sank to the ground.
A bird in a tree looked down and it said,
'Dear lion, I'm afraid that what you hear
Is simply the voice of your lion-sized fear.'

Brian Patten

The Crunch

The lion and his tamer
They had a little tiff,
For the lion limped too lamely,—
The bars had bored him stiff.

No need to crack your whip, Sir!
Said the lion then irate:
No need to snap my head off,
Said the tamer—but too late.

Gerda Mayer

Noah's Ark

It began
When God popped His head
Through the clouds and said:

'Oh you wicked, wicked children
What a mess this place is in
All the violence and corruption
It really is a sin.

I turn my back for five aeons
(For I've other work to do)
Keeping the universe tidy
And I get no thanks from you.

You've grown selfish and conceited
Your manners are a disgrace
You come and go just as you please
You'd think you owned the place.

A telling-off's not good enough
You've grown too big for your flesh
So I think I'll wash my hands of you
And start again afresh.'

He turned full on the tap in the sky
Then picked out the one good man
Pure of heart and strong in arm
To carry out his plan: Noah.

'What I need,' explained God
'Is an arkwright to build an ark, right away.'
Said Noah, 'If I can sir.'
'Of course you can, now get stuck in
I won't take Noah for an answer.'

'I want a boat three storeys high
Aboard which you will bring
Not only your wife and family
But two of every living thing.'

'Even spiders?' asked Noah
(who didn't really like them)
'Even spiders,' said God
(who didn't either).

'Cats and dogs and elephants
Slugs, leopards and lice
Giraffes and armadilloes
Buffaloes, bed bugs and mice.

Antelopes, ants and anteaters
(though keep those last two apart)
Bears from Koala to Grizzly
Horses from Racing to Cart.

Fish will be able to fend for themselves
And besides, a wooden ark
Is not the sort of place to keep
A whale or an angry shark.

And don't forget our feathered friends
For they'll have nowhere to nest.
But vermin will determine
Their own survival best.

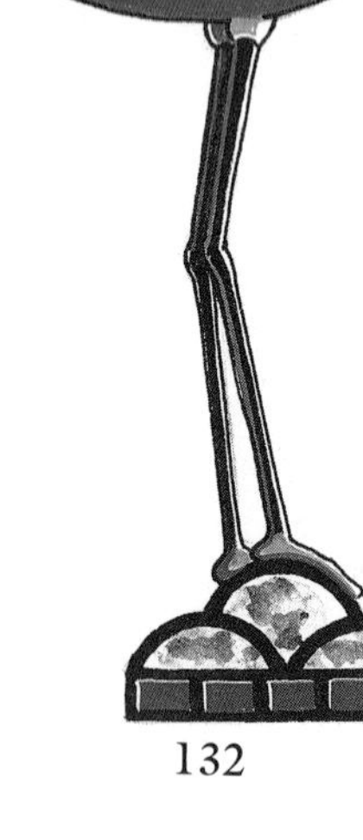

Flies, maggots and bluebottles
Mosquitoes and stingers that bite
Will live off the dead and dying
So they'll make out all right.

That seems to be all for now, Noah
The rest is up to you
I'll see you again in forty days
Until then God Bless and Adieu.'

He disappeared in a clap of thunder
(Either that or he banged the door)
And the wind in a rage broke out of its cage
With an earth-splintering roar.

And no sooner was everyone aboard
Than the Ark gave a mighty shudder
And would have been crushed by the onrushing waves
Had Noah not been at the rudder.

Right craftily he steered the craft
As if to the mariner born
Through seas as high as a Cyclop's eye
And cold as the devil's spawn.

And it rained, and it rained
And it rained again
And it rained, and it rained
And it rained, and then . . .
. . . drip . . .
. . . drop . . .
. . . the last . . .
. . . drip dropped . . .
. . to a . . .
. . . stop.

Noah at the helm was overwhelmed
For both cargo and crew were unharmed
Then the wind turned nasty and held its breath
So the Ark became becalmed.

Hither and thither it drifted
Like an aimless piece of jetsam
'Food's running out,' cried Mrs Noah
'We'll perish if we don't get some.'

'Maybe God's gone and forgotten us
We're alone in the world and forsaken
He surely won't miss one little pig
Shall I grill a few rashers of bacon?'

'Naughty, naughty!' said Noah sternly
(For it was the stern that he was stood in)
'I'm ravenous, but bring me a raven
I've an idea and I think it's a good 'un.'

As good as his word, he let loose the bird
'Go spy out for land,' he commanded
But in less than a week, it was back with its beak
Completely (so to speak) empty-handed!

Next he coaxed from its lovenest a dove
'We're depending on you,' he confided
Then gave it to the air like an unwrapped gift
Of white paper, that far away glided.

Then the Ark sat about with its heart in its mouth
With nothing to do but wait
So Mrs Noah organized organized games
To keep animal minds off their fate.

Until one morn when all seemed lost
The dove in the heavens was seen
To the Ark, like an archangel it arrowed
Bearing good tidings of green.

'Praised be the Lord,' cried Noah
(and Mrs Noah cried too)
And all God's creatures gave their thanks
(even spiders, to give them their due).

Then God sent a quartet of rainbows
Radiating one from each side
To the four corners of the earth
Where they journeyed and multiplied.

And as Noah set off down the mountain
To be a simple farmer again
A voice thundered: 'Nice work there sunshine.'
Here endeth the story. Amen.

Roger McGough

The Pelican Chorus

King and Queen of the Pelicans we;
No other Birds so grand we see!
None but we have feet like fins!
With lovely leathery throats and chins!
    Ploffskin, Pluffskin, Pelican jee,
    We think no Birds so happy as we!
    Plumpskin, Ploshkin, Pelican jill,
    We think so then, and we thought so still!

We live on the Nile. The Nile we love.
By night we sleep on the cliffs above;
By day we fish, and at eve we stand
On long bare islands of yellow sand.
And when the sun sinks slowly down
And the great rock walls grow dark and brown,
Where the purple river rolls fast and dim
And the Ivory Ibis starlike skim,
Wing to wing we dance around,—
Stamping our feet with a flumpy sound,—
Opening our mouths as Pelicans ought,
And this is the song we nightly snort:—
    Ploffskin, Pluffskin, Pelican jee,
    We think no Birds so happy as we!
    Plumpskin, Ploshskin, Pelican jill,
    We think so then, and we thought so still!

Last year came out our Daughter, Dell;
And all the Birds received her well.
To do her honour, a feast we made
For every bird that can swim or wade.
Herons and Gulls, and Cormorants black,
Cranes, and Flamingoes with scarlet back,
Plovers and Storks, and Geese in clouds,
Swans and Dilberry Ducks in crowds.
Thousands of Birds in wondrous flight!
They ate and drank and danced all night,
And echoing back from the rocks you heard
Multitude-echoes from Bird and Bird,—

Ploffskin, Pluffskin, Pelican jee,
We think no Birds so happy as we!
Plumpskin, Ploshkin, Pelican jill,
We think so then, and we thought so still!

Yes, they came; and among the rest,
The King of the Cranes all grandly dressed.
Such a lovely tail! Its feathers float
Between the ends of his blue dress-coat;
With pea-green trousers all so neat,
And a delicate frill to hide his feet,—
(For though no one speaks of it, every one knows,
He has got no webs between his toes!)
As soon as he saw our Daughter, Dell,
In violent love that Crane King fell,—
On seeing her waddling form so fair,
With a wreath of shrimps in her short white hair.
And before the end of the next long day,
Our Dell had given her heart away;
For the King of the Cranes had won that heart,
With a Crocodile's egg and a large fish-tart.
She vowed to marry the King of the Cranes,
Leaving the Nile for stranger plains;
And away they flew in a gathering crowd
Of endless birds in a lengthening cloud.

Ploffskin, Pluffskin, Pelican jee,
We think no Birds so happy as we!
Plumpskin, Ploshkin, Pelican jill,
We think so then, and we thought so still!

And far away in the twilight sky,
We heard them singing a lessening cry,—
Farther and farther till out of sight,
And we stood alone in the silent night!
Often since, in the nights of June,
We sit on the sand and watch the moon;—
She has gone to the great Gromboolian plain,
And we probably never shall meet again!
Oft, in the long still nights of June,
We sit on the rocks and watch the moon;—
She dwells by the streams of the Chankly Bore,
And we probably never shall see her more.
    Ploffskin, Pluffskin, Pelican jee,
    We think no Birds so happy as we!
    Plumpskin, Ploshkin, Pelican jill,
    We think so then, and we thought so still!

Edward Lear

The Huntsman

The story is based on a folk tale heard by
the author in Kenya in 1944.

Kagwa hunted the lion,
   Through bush and forest went his spear.
One day he found the skull of a man
   And said to it, 'How did you come here?'
The skull opened its mouth and said
   'Talking brought me here.'

Kagwa hurried home;
   Went to the king's chair and spoke:
'In the forest I found a talking skull.'
   The king was silent. Then he said slowly
'Never since I was born of my mother
   Have I seen or heard of a skull which spoke.'

The king called out his guards:
   'Two of you now go with him
And find this talking skull:
   But if his tale is a lie
And the skull speaks no word,
   This Kagwa himself must die.'

They rode into the forest:
   For days and nights they found nothing.
At last they saw the skull; Kagwa
   Said to it 'How did you come here?'
The skull said nothing. Kagwa implored,
   But the skull said nothing.

The guards said 'Kneel down.'
   They killed him with sword and spear.
Then the skull opened its mouth;
   'Huntsman, how did you come here?'
And the dead man answered
   'Talking brought me here.'

Edward Lowbury

Childe Maurice

Childe Maurice hunted the Silver Wood,
He whistled and he sung:
'I think I see the woman yonder
That I have loved so long.'

He called to his little man John,
'You do not see what I see;
For yonder I see the very first woman
That ever lovèd me.'

'Here is a glove, a glove,' he says,
'Lined all with fur it is;
Bid her to come to Silver Wood
To speak with Childe Maurice.'

'And here is a ring, a ring,' he says,
'A ring of the precious stone:
He prays her come to Silver Wood
And ask the leave of none.'

'Well do I love your errand, master,
But better I love my life.
Would you have me go to John Steward's castle,
To tryst away his wife?'

'Do not I give you meat?' he says,
'Do not I give you fee?
How dare you stop my errand
When that I bid you flee?'

When the lad came to John Steward's castle,
He ran right through the gate
Until he came to the high, high hall
Where the company sat at meat.

'Here is a glove, my lady,' said he,
'Lined all with fur it is;
It says you're to come to Silver Wood
And speak with Childe Maurice.

'And here is a ring, a ring of gold,
Set with the precious stone:
It prays you to come to Silver Wood
And ask the leave of none.'

Out then spake the wily nurse,
A wily woman was she,
'If this be come from Childe Maurice
It's dearly welcome to me.'

'Thou liest, thou liest, thou wily nurse,
So loud as I hear thee lie!
I brought it to John Steward's lady,
And I trow thou be not she.'

Then up and rose him John Steward,
And an angry man was he:
'Did I think there was a lord in the world
My lady loved but me!'

He dressed himself in his lady's gown,
Her mantle and her hood;
But a little brown sword hung down by his knee,
And he rode to Silver Wood.

Childe Maurice sat in Silver Wood,
He whistled and he sung,
'I think I see the woman coming
That I have loved so long.'

But then stood up Childe Maurice
His mother to help from horse:
'O alas, alas!' says Childe Maurice,
'My mother was never so gross!'

'No wonder, no wonder,' John Steward he said,
'My lady loved thee well,
For the fairest part of my body
Is blacker than thy heel.'

John Steward took the little brown sword
    That hung low down by his knee;
He has cut the head off Childe Maurice
    And the body put on a tree.

And when he came to his lady—
    Looked over the castle-wall—
He threw the head into her lap,
    Saying, 'Lady, take the ball!'

Says, 'Dost thou know Childe Maurice' head,
    When that thou dost it see?
Now lap it soft, and kiss it oft,
    For thou loved'st him better than me.'

But when she looked on Childe Maurice' head,
    She ne'er spake words but three:
'I never bare no child but one,
    And you have slain him, trulye.'

'I got him in my mother's bower
    With secret sin and shame;
I brought him up in the good greenwood
    Under the dew and rain.'

And she has taken her Childe Maurice
    And kissed him, mouth and chin:
'O better I loved my Childe Maurice
    Than all my royal kin!'

'Woe be to thee!' John Steward he said,
    And a woe, woe man was he;
'For if you had told me he was your son
    He had never been slain by me.'

Says, 'Wicked be my merry men all,
    I gave meat, drink and cloth!
But could they not have holden me
    When I was in all that wrath?'

Traditional

Excelsior

The shades of night were falling fast,
As through an Alpine village passed
A youth, who bore, 'mid snow and ice,
A banner with the strange device,
    Excelsior!

His brow was sad; his eye beneath
Flashed like a faulchion from its sheath,
And like a silver clarion rung
The accents of that unknown tongue,
    Excelsior!

In happy homes he saw the light
Of household fires gleam warm and bright;
Above, the spectral glaciers shone,
And from his lips escaped a groan,
    Excelsior!

'Try not the Pass!' the old man said,
'Dark lowers the tempest overhead,
The roaring torrent is deep and wide!'
And loud that clarion voice replied,
  Excelsior!

'O stay!' the maiden said, 'and rest
Thy weary head upon this breast!'
A tear stood in his bright blue eye,
But still he answered, with a sigh,
  Excelsior!

'Beware the pine-tree's withered branch!
Beware the awful avalanche!'
This was the peasant's last goodnight!
A voice replied, far up the height,
  Excelsior!

At break of day, as heavenward
The pious monks of Saint Bernard
Uttered the oft-repeated prayer,
A voice cried through the startled air,
  Excelsior!

A traveller, by the faithful hound,
Half-buried in the snow, was found,
Still grasping in his hand of ice
That banner, with the strange device
  Excelsior!

There, in the twilight cold and grey,
Lifeless, but beautiful, he lay,
And from the sky, serene, and far,
A voice fell, like a falling star,
  Excelsior!

Henry Wadsworth Longfellow

A Legend of the Northland

Away, away in the Northland,
Where the hours of the day are few,
And the nights are so long in winter
That they cannot sleep them through;

Where they harness the swift reindeer
To the sledges, when it snows;
And the children look like bear's cubs
In their funny, furry clothes:

They tell them a curious story—
I don't believe 'tis true;
And yet you may learn a lesson
If I tell the tale to you.

Once, when the good Saint Peter
Lived in the world below,
And walked about it, preaching,
Just as he did, you know,

He came to the door of a cottage,
In travelling round the earth,
Where a little woman was making cakes,
And baking them on the hearth;

And being faint with fasting,
For the day was almost done,
He asked her, from her store of cakes,
To give him a single one.

So she made a very little cake,
But as it baking lay,
She looked at it, and thought it seemed
Too large to give away.

Therefore she kneaded another,
And still a smaller one;
But it looked, when she turned it over,
As large as the first had done.

Then she took a tiny scrap of dough,
And rolled and rolled it flat;
And baked it thin as a wafer—
But she couldn't part with that.

For she said, 'My cakes that seem too small
When I eat of them myself
Are yet too large to give away.'
So she put them on the shelf.

Then good Saint Peter grew angry,
For he was hungry and faint;
And surely such a woman,
Was enough to provoke a saint.

And he said, 'You are far too selfish
To dwell in a human form,
To have both food and shelter,
And fire to keep you warm.

'Now, you shall build as the birds do,
And shall get your scanty food
By boring, and boring, and boring,
All day in the hard, dry wood.'

Then up she went through the chimney,
Never speaking a word,
And out of the top flew a woodpecker,
For she was changed to a bird.

She had a scarlet cap on her head,
And that was left the same,
But all the rest of her clothes were burned
Black as a coal in the flame.

And every country schoolboy
Has seen her in the wood,
Where she lives in the trees till this very day,
Boring and boring for food.

Phoebe Cary

The Story of Canobie Dick

A tale from the Borders

A bad day at the market for Canobie Dick,
gone there to sell his horses, and no luck:

'Just have to take them home again—and it's late.
Still, there's always the short cut,
west by the hills—'

'The Eildon Hills? I wouldn't! Not
this time of night!'

'Afraid of the fairies, eh?' says Dick,
'I'm not such a fool!'

His two black horses
pick their way up through the gorse
under a pinch of stars, and nothing between but the wind
whirling down Scotland, and him thinking of his empty purse,
when there's a low voice speaks out, right in his path:
Are you selling your horses, man? I'll buy them both.

'Who's there?'

Only a great black rock, crouched like a hare.

'Come on, show yourself! Ghost or devil I don't care,
so long as your money's good!'

The wind's dropped.
A slice of darkness moves out of the dark rock,
and he stares
into the tall shape of a man, what face he has
hidden by a hood—
with a heap of coins held out in a thin hand.

'This is old-fashioned money!'
Dick laughs. 'Where'd you find it?
Looks like it's spent
years in the ground . . .
But it's gold right enough,
I'll take it—
the horses are yours!'

I'll need more

Like the creaking of a forgotten
door in a lost house, that voice
that holds him still—

If you have more to sell, then bring them
here to me at midnight.

Says Dick '—I will!'

One night, another—
but where is he taking them?
No house for miles,
it's as if the hill swallows him!
Ghost he may be, but my horses are real—
there's no trace of them!

So the last night Dick says, 'Look,
it's a dreary time and place for buying and selling—
maybe, could I warm myself a while at your dwelling?'

The old man looks at him,
the starlight catching
stars from his hidden eyes.
He sighs:

Very well, if you will.
But I warn you—remember—
if your courage should fail
you'll be sorry for ever.
Follow me then, if you dare.

There's a crack in the rock
dark as death's door.
'That's odd,' says Dick,
'I never noticed it there before?'
The old man steps through,
Dick stumbles behind him
into the earth and down.
Hand and foot feel
a chill tunnel of stone—
You may still turn back,
says the old man, turning.
Says Dick, 'Lead on!'

And then, round a corner,
the walls open, the roof soars,
and a sudden flare
of torches blazing there,
but white, like moonlight, lays bare
a huge hall—
a hall
fit for a palace, buried in the hill!

That's where his horses are!
Stabled each side of it
horse after horse,
every one of them black as water in a deep well—
and as still . . .

But it's not the horses makes him tremble.
Right down the hall runs a stone table,
and sprawled on benches on either side lie
mighty warriors, knights in armour—
coal-black their armour, their matted beards—
heads lolled, mouths gaping,
and not dead:
sleeping.
The air thrums
with the harsh tides of their dreams . . .

And between them,
there, on the table,
nothing for eating,
nothing for drinking,
nothing at all
but a single sword in a scabbard
and a twisted hunting horn.

Run, his heart cries to him, *run!*

He looks round—but the old man's
lifted back his hood
on a face grey as weathered wood
in a stone wall.
Heavy now his words fall:

So it is written:
he who shall sound that horn and draw that sword,
if his heart fail not, he shall be King of Britain.
So speaks the tongue that cannot lie.

King, says Dick, King—me?
I'm thinking long ago
I saw all this in a nightmare—
but what did I do?

What did I do?
What should I do?

But already his feet
are stepping him forward:
he touches the table,
he touches the sword,
he jolts it, it clatters—
only a little,
but a sleeper mutters,
a sleeper stirs with a groan—
Dick grabs the horn!

Shaking like grass
he can scarcely press
its cold little mouth to his own—
King, king—this must be a dream?
He shuts his eyes.
He blows.

Thin as the scream of a rabbit
the cracked note runs round the walls—

And at once in a crash of stone like thunder,
a whinny of wild-eyed horses, all
the warriors leap to their feet with a shout
the length of the hall,
swords drawn, and drawn at him!
And a great voice above the din cries out:

WOE TO THE COWARD THAT EVER HE WAS BORN
WHO DID NOT DRAW THE SWORD BEFORE HE BLEW THE HORN!

But the rage in the air has snatched him away like paper,
black and white tumbling round him, howling to hurl him
in a whirlwind back through the tunnel, from stone to stone
and out, on to the moor, into the night, alone
where he heard the door in the rock clash closed for ever.

It was there on the dewy grass the shepherds found him
and no more breath beyond this story he told them.

Libby Houston

How They Brought the Good News from Ghent to Aix

I sprang to the stirrup, and Joris, and he;
I galloped, Dirck galloped, we galloped all three;
'Good speed!' cried the watch, as the gate-bolts undrew;
'Speed!' echoed the wall to us galloping through;
Behind shut the postern, the lights sank to rest,
And into the midnight we galloped abreast.

Not a word to each other; we kept the great pace
Neck by neck, stride by stride, never changing our place;
I turned in my saddle and made its girths tight,
Then shortened each stirrup, and set the pique right,
Rebuckled the cheek-strap, chained slacker the bit,
Nor galloped less steadily Roland a whit.

'Twas moonset at starting; but while we drew near
Lokeren, the cocks crew and twilight dawned clear;
At Boom, a great yellow star came out to see;
At Duffeld, 'twas morning as plain as could be;
And from Mecheln church-steeple we heard the half-chime
So Joris broke silence with, 'Yet there is time!'

At Aerschot, up leaped of a sudden the sun,
And against him the cattle stood black every one,
To stare thro' the mist at us galloping past,
And I saw my stout galloper Roland at last,
With resolute shoulders, each butting away
The haze, as some bluff river headland its spray.

And his low head and crest, just one sharp ear bent back
For my voice, and the other pricked out on his track;
And one eye's black intelligence—ever that glance
O'er its white edge at me, his own master, askance!
And the thick heavy spume-flakes which aye and anon
His fierce lips shook upwards in galloping on.

By Hasselt, Dirck groaned; and, cried Joris, 'Stay spur!!
Your Roos galloped bravely, the fault's not in her,
We'll remember at Aix'—for one heard the quick wheeze
Of her chest, saw the stretched neck and staggering knees,
And sunk tail, and horrible heave of the flank,
As down on her haunches she shuddered and sank.

So we were left galloping, Joris and I,
Past Looz and past Tongres, no cloud in the sky;
The broad sun above laughed a pitiless laugh,
'Neath our feet broke the brittle bright stubble like chaff;
Till over by Dalhem a dome-spire sprang white,
And 'Gallop,' gasped Joris, 'for Aix is in sight!'

'How they'll greet us!'—and all in a moment his roan
Rolled neck and crop over, lay dead as a stone;
And there was my Roland to bear the whole weight
Of the news which alone could save Aix from her fate,
With his nostrils like pits full of blood to the brim,
And with circles of red for his eye-sockets' rim.

Then I cast loose my buffcoat, each holster let fall,
Shook off both my jack-boots, let go belt and all,
Stood up in the stirrup, leaned, patted his ear,
Called my Roland his pet-name, my horse without peer;
Clapped my hands, laughed and sang, any noise, bad or good,
Till at length into Aix Roland galloped and stood.

And all I remember is, friends flocking round
As I sat with his head 'twixt my knees on the ground;
And no voice but was praising this Roland of mine,
As I poured down his throat our last measure of wine,
Which (the burgesses voted by common consent)
Was no more than his due who brought good news from Ghent.

Robert Browning

The Highwayman

Part I

The wind was a torrent of darkness among the gusty trees,
The moon was a ghostly galleon tossed upon cloudy seas.
The road was a ribbon of moonlight over the purple moor,
And the highwayman came riding—
        Riding—riding—
The highwayman came riding, up to the old inn-door.

He'd a French cocked-hat on his forehead, a bunch of lace at his chin,
A coat of the claret velvet, and breeches of brown doe-skin.
They fitted with never a wrinkle. His boots were up to the thigh.
And he rode with a jewelled twinkle,
        His pistol butts a-twinkle,
His rapier hilt a-twinkle, under the jewelled sky.

Over the cobbles he clattered and clashed in the dark inn-yard.
He tapped with his whip on the shutters, but all was locked and barred.
He whistled a tune to the window, and who should be waiting there
But the landlord's black-eyed daughter,
        Bess, the landlord's daughter,
Plaiting a dark red love-knot into her long black hair.

And dark in the dark old inn-yard a stable-wicket creaked
Where Tim the ostler listened. His face was white and peaked.
His eyes were hollows of madness, his hair like mouldy hay,
But he loved the landlord's daughter,
        The landlord's red-lipped daughter.
Dumb as a dog he listened, and he heard the robber say—

'One kiss, my bonny sweetheart, I'm after a prize to-night,
But I shall be back with the yellow gold before the morning light;
Yet, if they press me sharply, and harry me through the day,
Then look for me by moonlight,
        Watch for me by moonlight,
I'll come to thee by moonlight, though hell should bar the way.'

He rose upright in the stirrups. He scarce could reach her hand,
But she loosened her hair i' the casement. His face burnt like a brand
As the black cascade of perfume came tumbling over his breast;
And he kissed its waves in the moonlight,
        (Oh, sweet black waves in the moonlight!)
Then he tugged at his rein in the moonlight, and galloped away to the west.

Part II

He did not come in the dawning. He did not come at noon;
And out o' the tawny sunset, before the rise o' the moon,
When the road was a gipsy's ribbon, looping the purple moor,
A red-coat troop came marching—
        Marching—marching—
King George's men came marching, up to the old inn-door.

They said no word to the landlord. They drank his ale instead.
But they gagged his daughter, and bound her, to the foot of her narrow bed.
Two of them knelt at her casement, with muskets at their side!
There was death at every window;
        And hell at one dark window;
For Bess could see, through her casement, the road that *he* would ride.

They had tied her up to attention, with many a sniggering jest.
They had bound a musket beside her, with the muzzle beneath her breast!
'Now, keep good watch!' and they kissed her.
        She heard the dead man say—
Look for me by moonlight;
        Watch for me by moonlight;
I'll come to thee by moonlight, though hell should bar the way!

She twisted her hands behind her; but all the knots held good!
She writhed her hands till her fingers were wet with sweat or blood!
They stretched and strained in the darkness, and the hours crawled by like years,
Till now, on the stroke of midnight,
        Cold, on the stroke of midnight,
The tip of one finger touched it! The trigger at least was hers!

The tip of one finger touched it. She strove no more for the rest.
Up, she stood up to attention, with the muzzle beneath her breast.
She would not risk their hearing; she would not strive again;
For the road lay bare in the moonlight;
        Blank and bare in the moonlight;
And the blood of her veins, in the moonlight, throbbed to her love's refrain.

Tlot-tlot; tlot-tlot! Had they heard it? The horse-hoofs ringing clear;
Tlot-tlot, tlot-tlot, in the distance! Were they deaf that they did not hear?
Down the ribbon of moonlight, over the brow of the hill,
The highwayman came riding,
        Riding, riding!
The red-coats looked to their priming! She stood up, straight and still.

Tlot-tlot, in the frosty silence! *Tlot-tlot,* in the echoing night!
Nearer he came and nearer. Her face was like a light.
Her eyes grew wide for a moment; she drew one last deep breath,
Then her finger moved in the moonlight,
        Her musket shattered the moonlight,
Shattered her breast in the moonlight and warned him—with her death.

He turned. He spurred to the west; he did not know who stood
Bowed, with her head o'er the musket, drenched with her own red blood!
Not till the dawn he heard it, and his face grew grey to hear
How Bess, the landlord's daughter,
        The landlord's black-eyed daughter,
Had watched for her love in the moonlight, and died in the darkness there.

Back, he spurred like a madman, shouting a curse to the sky,
With the white road smoking behind him and his rapier brandished high.
Blood-red were his spurs i' the golden noon; wine-red was his velvet coat;
When they shot him down on the highway,
        Down like a dog on the highway,
And he lay in his blood on the highway, with the bunch of lace at his throat.

And still of a winter's night, they say, when the wind is in the trees,
When the moon is a ghostly galleon tossed upon cloudy seas,
When the road is a ribbon of moonlight over the purple moor,
A highwayman comes riding—
        Riding—riding—
A highwayman comes riding, up to the old inn-door.

Over the cobbles he clatters and clangs in the dark inn-yard.
And he taps with his whip on the shutters, but all is locked and barred.
He whistles a tune to the window, and who should be waiting there
But the landlord's black-eyed daughter,
Bess, the landlord's daughter,
Plaiting a dark red love-knot into her long black hair.

Alfred Noyes

The Man in the Moon Stayed up Too Late

There is an inn, a merry old inn
  beneath an old grey hill,
And there they brew a beer so brown
That the Man in the Moon himself came down
  one night to drink his fill.

The ostler has a tipsy cat
  that plays a five-stringed fiddle;
And up and down he runs his bow,
Now squeaking high, now purring low,
  now sawing in the middle.

The landlord keeps a little dog
  that is mighty fond of jokes;
When there's good cheer among the guests,
He cocks an ear at all the jests
  and laughs until he chokes.

They also keep a hornèd cow
  as proud as any queen;
But music turns her head like ale,
And makes her wave her tufted tail
  and dance upon the green.

And O! the row of silver dishes
  and the store of silver spoons!
For Sunday there's a special pair,
And these they polish up with care
  on Saturday afternoons.

The Man in the Moon was drinking deep,
and the cat began to wail;
A dish and a spoon on the table danced,
The cow in the garden madly pranced,
and the little dog chased his tail.

The Man in the Moon took another mug
and then rolled beneath his chair.
And there he dozed and dreamed of ale,
Till in the sky the stars were pale,
and dawn was in the air.

The ostler said to his tipsy cat:
'The white horses of the Moon,
They neigh and champ their silver bits;
But their master's been and drowned his wits,
and the Sun'll be rising soon!'

So the cat on his fiddle played hey-diddle-diddle,
a jig that would wake the dead;
He squeaked and sawed and quickened the tune,
While the landlord shook the Man in the Moon:
'It's after three!' he said.

They rolled the Man slowly up the hill
and bundled him into the Moon,
While his horses galloped up in rear,
And the cow came capering like a deer,
and a dish ran up with a spoon.

Now quicker the fiddle went deedle-dum-diddle;
the dog began to roar,
The cow and the horses stood on their heads;
The guests all bounded from their beds
and danced upon the floor.

With a ping and a pong the fiddle-strings broke!
the cow jumped over the Moon,
And the little dog laughed to see such fun,
And the Saturday dish went off at a run
With the silver Sunday spoon.

The round Moon rolled behind the hill,
as the Sun raised up her head.
She hardly believed her fiery eyes;
For though it was day, to her surprise
they all went back to bed!

J. R. R. Tolkien

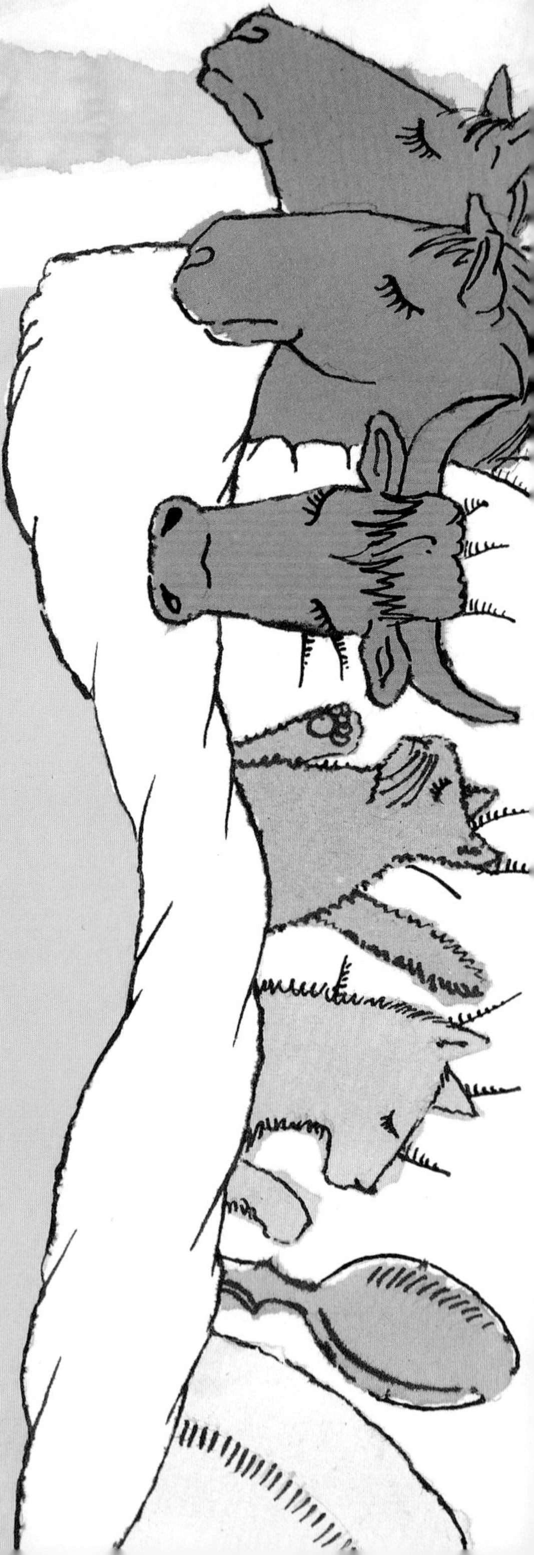

Switch on the Night

Once there was a little boy
who didn't like the Night.

He liked
lanterns and lamps
and
torches and tapers
and
beacons and bonfires
and
flashlights and flares.
But he didn't like the Night.

He didn't like light switches at all.
Because light switches turned off
the yellow lamps
the green lamps
the white lamps
the hall lights
the house lights
the lights in all the rooms.
He wouldn't touch a light switch.

And he wouldn't go out to play
after dark.
He was very lonely.
And unhappy.
For he saw, from his window,
the other children playing
on the summer-night lawns.
In and out of the dark and
lamplight ran the children . . .
happily.

But where was our little boy?
Up in his room.
With his lanterns and lamps
and flashlights
and candles and chandeliers.
All by himself.

He liked only the sun.
The yellow sun.
He didn't like
the Night.

When it was time for Mother and Father
to walk around switching off all the
lights . . .
One by one.

One by one.
The porch lights
the parlour lights
the pale lights
the pink lights
the pantry lights
and stairs lights . . .
Then the little boy hid in his bed.

Late at night
his was the only room
with a light
in all the town.

And then one night
With his father away on a trip
And his mother gone to bed early,
The little boy wandered alone,
All alone through the house.

My, how he had the lights blazing!
the parlour lights
and porch lights
the pantry lights
the pale lights
the pink lights
the hall lights
the kitchen lights
even the *attic* lights!
The house looked like it was on fire!

But still the little boy was alone.
While the other children played
on the night lawns.
Laughing.
Far away.

All of a sudden he heard
a rap at a window!
Something dark was there.
A knock on the screen door.
Something dark was there!
A tap at the back porch.
Something dark was there!

And all of a sudden someone said 'Hello!'
And a little girl stood there in the middle of
the white lights, the bright lights,
the hall lights, the small lights,
the yellow lights, the mellow lights.

'My name is Dark,' she said.
And she had dark hair,
and dark eyes,
and wore a dark dress
and dark shoes.
But her face was as white as the moon.
And the light in her eyes
shone like white stars.

'You're lonely,' she said.

'Think what you're missing!
Have you ever thought of
switching on the crickets,
switching on the frogs,
switching on the stars,
and the great big white moon?'

'No,' said the little boy.

'Well, try it,' said Dark.
And they did.

They climbed up and down stairs,
switching on the Night.
Switching on the dark.
Letting the Night live in every room.
Like a frog.
Or a cricket.
Or a star.
Or a moon.

And they switched on the crickets.
And they switched on the frogs.
And they switched on the white, ice-cream moon.

'Oh, I like this!' said the little boy.
'Can I switch on the Night always?'

'Of course!' said Dark, the little girl.
And then she vanished.

And now the little boy is very happy.
He likes the Night.
Now he has a Night-switch instead of a light-switch!

He likes switches now.
He threw away his candles
and flashlights
and lamplights.
And any night in summer that you wish
you can see him
switching on the white moon,
switching on the red stars,
switching on the blue stars,
the green stars, the light stars,
the white stars,
switching on the frogs, the crickets, and Night.

And running in the dark, on the lawns,
with the happy children . . .
Laughing.

Ray Bradbury

Themes

Animals:
Combinations
Jim Who Ran Away from his Nurse
Jonah and the Whale
Noah's Ark
The Ambitious Ant
The Centipede
The Complacent Tortoise
The Crunch
The Jackdaw of Rheims
The Lion and the Echo
The Little Turtle
The Owl and the Pussy-Cat
The Pelican Chorus
The Unicorn

Journeys:
A Speck Speaks
Excelsior
How They Brought the Good News
Journey
Sir Patrick Spens
The Owl and the Pussy-Cat

Meetings:
Ballad
Childe Maurice
Fairy Story
La Belle Dame Sans Merci
Meet-on-the-Road
The Huntsman
The Old Woman and the Sandwiches
The Story of Canobie Dick

Monsters:
Jabberwocky
The Hairy Toe
The Malfeasance
Welsh Incident

Mystery:
A Legend of the Northland
As Lucy Went A-Walking
By St Thomas Water
Fairy Story
Mary Celeste
Humpty Dumpty's Recitation
Jabberwocky
Life Story
Lost
One Winter Night in August
The Man in the Moon Stayed up Too Late
The Nose
The Sad Story of a Little Boy That Cried
The Walrus and the Carpenter
Three Wise Old Women

Night:
Switch on the Night
The Listeners
The Man in the Moon Stayed up Too Late

Nonsense:
Combinations
Good Taste
Here is the News

People:
A Legend of the Northland
Annabel Lee
Annabell and the Witches
Bishop Hatto
Childe Maurice
Count Carrots
Jim Who Ran Away from his Nurse
La Belle Dame Sans Merci
Mr Tom Narrow
P. C. Plod versus the Dale St Dogstrangler
The Highwayman
The Huntsman
The Lady of Shalott
The Pied Piper of Hamelin
Three Wise Old Women
Uncle Alfred's Long Jump
What Happened to Miss Frugle

Sea:
A Speck Speaks
Jonah and the Whale
Mary Celeste
Noah's Ark
Sir Patrick Spens
The Forsaken Merman
The Inchcape Rock

Telling Stories:
After Ever Happily
Life Story
Storytime
Welsh Incident
Meet-on-the-Road
Merlin and the Snake's Egg
The Apple-Tree Man
The Lady of Shalott
The Listeners
The Story of Canobie Dick
What has Happened to Lulu?

Index of authors

The artists

The illustrations are by:

Shirley Barker (pp 45, 46/47, 109, 128, 148); **Jenny Brackley** (pp 23, 24, 63, 64/65, 66/67, 70/71, 100/101, 102/103, 105, 114/5, 155, 157); **Frances Cony** (pp 72/3); **Caroline Crossland** (pp 25, 26, 74/5, 99); **Lydia Evans** (pp 13, 20/21, 96/7, 112/113, 122); **Robina Green** (pp 8/9, 34/5, 36/7, 61, 62, 106/7, 169, 170/171); **Tudor Humphries** (pp 14/15, 48/9, 50/51, 150/151, 152/3); **Valerie McBride** (pp 19, 31, 116/7, 118/9, 120/121, 131, 132/3, 134/5); **Lyn O'Neill** (p 39); **Nicki Palin** (pp 10/11, 43, 53, 54/5, 137, 138/9, 145, 146); **Paul Thomas** (p 41); **Martin White** (pp 6/7, 16/17, 57, 58, 77, 78/9, 80, 110/111, 164/5, 166); **Freire Wright** (pp 28, 83, 84, 86, 124/5, 127, 141, 143)

The editors and publisher are grateful to Mrs Renate Keeping for kind permission to reproduce illustrations by the late Charles Keeping (on pp 88/9, 90/91, 92/3, 94, 159, 160/161, 163).

The jacket illustration is by Jenny Brackley

Index of titles and first lines

First lines are shown in italics

Acknowledgements

The editors and publisher are grateful for permission to include the following copyright material:

Hilaire Belloc: 'Jim and the Lion' from *Cautionary Tales for Children* (Gerald Duckworth & Co. Ltd.). Reprinted by permission of the Peters Fraser & Dunlop Group Ltd. **Alan Bold:** 'The Malfeasance'. Reprinted by permission of the author. **Ray Bradbury:** 'Switch on the Night'. © 1955 by Ray Bradbury, renewed 1983. Reprinted by permission of Don Congdon Associates, Inc., and Pantheon Books, a division of Random House, Inc. **Charles Causley:** 'What has Happened to Lulu?' from *Figgie Hobbin*, 'The Apple-Tree Man' from *Jack the Treacle Eater* and 'By St Thomas Water' from *Union Street* (all published by Macmillan). Reprinted by permission of David Higham Associates Ltd. **Walter de la Mare:** 'The Listeners' and 'As Lucy Went A-Walking' from *Complete Poems.* Reprinted by permission of The Literary Trustees of Walter de la Mare and The Society of Authors as their representative. **James Godden:** 'Lost' from *Slowly But Surely* (lyric only), copyright 1911 Keith Prowse Music Publishing Ltd. Reproduced by permission of EMI Music Publishing Ltd., and International Music Publications. **Mick Gowar:** 'Annabell and the Witches' from *Third Time Lucky* (Viking Kestrel 1988), copyright © Mick Gowar 1988. Reprinted by permission of Penguin Books Ltd., and Murray Pollinger. **Robert Graves:** 'Welsh Incident' from *Collected Poems* by Robert Graves. Copyright © 1975 by Robert Graves. Reprinted by permission of Oxford University Press, Inc. and A. P. Watt Ltd. on behalf of the Executors of the Estate of Robert Graves. **Mary Ann Hoberman:** 'Combinations' from *Bugs* (Viking). Copyright © 1976 by Mary Ann Hoberman. Reprinted by permission of Gina Maccoby Literary Agency. **Libby Houston:** 'The Old Woman and the Sandwiches', © Libby Houston 1971, first published in *Plain Clothes* (Allison & Busby); 'The Story of Canobie Dick', © Libby Houston, commissioned by BBC Schools Radio and first broadcast 1983, reprinted from *Necessity* (Slow Dancer Press), © 1988 by Libby Houston. Used by permission of the author. **X. J. Kennedy:** 'One Winter Night in August' from *One Winter Night in August*, copyright © 1975 by X. J. Kennedy. Reprinted by permission of Curtis Brown, Ltd. **Vachel Lindsay:** 'The Little Turtle' from *Collected Poems.* Copyright 1920 by Macmillan Publishing Company, renewed 1948 by Elizabeth C. Lindsay. Reprinted by permission of Macmillan Publishing Company. **Christopher Logue:** 'Good Taste' from *Ode to the Dodo.* Reprinted by permission of Jonathan Cape Ltd., and the author. **Edward Lowbury:** 'The Huntsman'. Reprinted by permission of the author. **Gerda Mayer:** 'Ballad', first published in *New Angles I*, ed. John Foster; 'Count Carrots' and 'The Crunch', both first published in *The Candy Floss Tree*, poems by Norman Nicholson, Gerda Mayer and Frank Flynn. All reprinted by permission of the author. **Roger McGough:** 'P. C. Plod versus the Dale St Dogstrangler' from *You Tell Me* (Viking Kestrel). Reprinted by permission of Peters Fraser & Dunlop Group Ltd. 'Noah's Ark' from *Waving at Trains.* Reprinted by permission of the Peters Fraser & Dunlop Group Ltd., and Jonathan Cape Ltd. **Adrian Mitchell:** 'A Speck Speaks' from *Nothingmas Day* (Allison and Busby). Reprinted by permission of W. H. Allen & Co. plc. **Judith Nicholls:** 'Storytime' and 'Mary Celeste' from *Midnight Forest.* Reprinted by permission of Faber & Faber Ltd; 'Journey', © 1989 Judith Nicholls. Reprinted by permission of the author. **Leslie Norris:** 'Merlin and the Snake's Egg'. Reprinted by permission of the author. **Alfred Noyes:** 'The Highwayman' from *Collected Poems.* Reprinted by permission of John Murray (Publishers) Ltd. **Gareth Owen:** 'Uncle Alfred's Long Jump' from *Bright Lights Blaze Out* (Oxford University Press) and 'Jonah and the Whale' from *Salford Road* (Kestrel). Reprinted by permission of Rogers Coleridge & White Ltd. **Brian Patten:** 'What Happened to Miss Frugle', 'The Complacent Tortoise' and 'The Lion and the Echo' from *Gargling With Jelly* (Viking Kestrel 1985), copyright © Brian Patten 1985. Reprinted by permission of the author, and Penguin Books Ltd. **James Reeves:** 'Mr Tom Narrow' from James Reeves: *The Complete Poems for Children* (Heinemann)/*The Wandering Moon and Other Poems* (Puffin Books). © James Reeves. Reprinted by permission of The James Reeves Estate. **E. V. Rieu:** 'The Unicorn' from *Cuckoo Calling* (Methuen). Reprinted by permission of Richard Rieu. **Michael Rosen:** 'Here is the News' from *Wouldn't You Like to Know.* Reprinted by permission of André Deutsch Ltd. **Ian Serraillier:** '"After Ever Happily" or The Princess and the Woodcutter' from *Happily Ever After* (Oxford University Press, 1963). Reprinted by permission of the author. **Iain Crichton Smith:** 'The Nose' from *In the Middle.* Reprinted by permission of Victor Gollancz Ltd. **Stevie Smith:** 'Fairy Story' from *The Collected Poems of Stevie Smith* (Penguin Modern Classics/New Directions). Copyright © 1972 by Stevie Smith. Reprinted by permission of James MacGibbon, Literary Executor & New Directions Publishing Corporation. **J. R. R. Tolkien:** excerpt from 'The Man in the Moon Stayed up Too Late' from *The Adventures of Tom Bombadil.* Copyright © 1962 by George Allen & Unwin Ltd. Reprinted by permission of Unwin Hyman Ltd., and Houghton Mifflin Co.

Oxford University Press, Walton Street, Oxford 0X2 6DP

Oxford New York Toronto
Delhi Bombay Calcutta Madras Karachi
Petaling Jaya Singapore Hong Kong Tokyo
Nairobi Dar es Salaam Cape Town
Melbourne Auckland

and associated companies in
Berlin Ibadan

Oxford is a trade mark of Oxford University Press

First published 1990
First published by Oxford in the United States 1990

Library of Congress Catalog Card Number: 89 – 043715

British Library Cataloguing in Publication Data
Oxford book of story poems.
1. Children's poetry in English, to 1980 – Anthologies
I. Harrison, Michael, 1939 II. Stuart-Clark, Christopher
821'.008'09282

ISBN 0 19 276087 4

Set by Pentacor Ltd, High Wycombe, Bucks
Printed in Singapore